Systems Thinking Basics

Systems Thinking Basics

From Concepts to Causal Loops

Virginia Anderson
AND Lauren Johnson

PEGASUS COMMUNICATIONS, INC.

Cambridge, Massachusetts

PEGASUS COMMUNICATIONS, INC.
PO Box 120   Kendall Square
Cambridge, MA 02142-0001

Web Site:	http://www.pegasuscom.com
Phone:	(617) 576-1231
Order Phone:	(800) 272-0945
Fax:	(617) 576-3114
Order Fax:	(800) 701-7083

Library of Congress Cataloging-in-Publication Data

Anderson, Virginia, 1947–
    Systems Thinking Basics: From Concepts to Causal Loops / by Virginia Anderson
and Lauren Johnson.
        p.   cm.
        ISBN 1-883823-12-9
1. Industrial management—Case studies—Problems, exercises, etc.
2. Systems analysis—Problems, exercises, etc. 3. Decision making—exercises, etc.
I. Johnson, Lauren.   II. Title.
HD30. 19. A53   1997
658.4'032—dc21   96-39762
                                                CIP

 Printed on recycled paper

CONTENTS

Why Use This Book?

Congratulations! By picking up this book, you've taken the first step in learning to use the powerful array of systems thinking tools. These tools offer a fresh, highly effective way to grasp the complexities of organizational life and to address the stubborn problems that often confront us in the business world. *Systems Thinking Basics: From Concepts to Causal Loops* is designed to help you discover the principles of systems and systems thinking and begin actually using systems thinking tools. There is a broad palette of systems thinking tools available; this book emphasizes two foundational devices: *behavior over time graphs* and *causal loop diagrams.*

Why Systems Thinking?

Why take time out of your busy schedule to read this book? Because systems thinking offers a valuable new perspective on our most persistent organizational problems and our role in them. One of the major points that systems thinking makes is that everything—and everyone—is interconnected in an infinitely complex network of systems. When we begin to see the world—and one another—through this "lens," we start seeing our circumstances in a new light, taking more responsibility for our own role in problems, and identifying more effective ways of addressing recurring difficulties. Systems thinking has a power and a potential that, once you've been introduced, are hard to resist. You'll learn more about the benefits of systems thinking in Section 2.

How to Use This Book

The world of systems and systems thinking marks a dramatic shift from the more linear, analytic way of thinking that many of us are used to. It's especially challenging to convey the abstract characteristics of systems in print. Because systems themselves are so complex and so interconnected with everything around them, it's also impossible to capture any one entire system on paper. For this reason, the tools and activities in this book are meant to offer primarily a glimpse into systems behavior.

We also hope you will see the book as the starting point to a long, learning-filled journey. Systems research and theories are constantly evolving, and the concepts and tools in this book offer only one of many methods for looking at the whole. When you finish the book, you can continue your adventure into exploring systems thinking tools by using some of the resources listed in Appendix F.

In writing *Systems Thinking Basics,* we not only needed to decide what tools to include; we had to choose a plan for sequencing the sections. We opted for a format that allows each section to build on the one that came before, because that arrangement seemed best suited for new systems thinkers. For this reason, we recommend that you read the sections in numerical order, although you are certainly free to choose a different order depending on your interests and your familiarity with the material. We also hope that you'll work as often as possible with others on the readings and activities—it's when people use systems thinking tools *together* that these tools are their most potent.

Your journey through *Systems Thinking Basics* will begin with an exploration of the unique characteristics of systems (Section 1), and then move on to a definition of systems thinking and an explanation of its importance, especially to organizational life (Section 2). In sections 3 and 4, you will have the opportunity to create your own diagrams of systemic structures, step by step. Section 5 then gives you a taste of complex systems. Finally, Section 6 answers the question, "What next?," by offering advice on practicing systems thinking in your day-to-day life.

About the Learning Activities

Mastering systems thinking tools requires practice with lots of real-life examples. Accordingly, each section in this book contains a wealth of illustrations from the business world. Most sections then present a series of learning activities that challenge you to apply your new knowledge. The learning activities can be done as self-study or in groups, but again, we encourage you to work in groups as much as possible. We invite you to share with each other your responses to the activities and the insights you gain as you work with the activities.

The learning activities range from reflection and discussion to graphing and diagramming, and are filled with case studies adapted from recent analyses published in leading business journals, including *The Systems Thinker*™ newsletter. There is also a section (Appendix A) that offers extra learning activities should you desire additional practice. For the majority of the learning activities, you can check your responses against the "Learning Activity Key Points and Suggested Responses" in Appendix B. Remember, however, that there is no one right way to describe and diagram a system. These activities, and the suggested responses, are meant to spark your imagination and serve as a starting point for you to think about systems.

Acknowledgments

We'd like to thank the following colleagues who kindly reviewed the manuscript for this book. Each of them offered valuable insights and thoughtful suggestions that greatly strengthened the book:

Turina Bakken, MAQIN

Polly Bashore, General Motors

Dave Buffenbarger, New Dimensions in Learning

Lou Ann Daly, Innovation Associates

Richard Karash, Speaker, Facilitator, Trainer, "Towards Learning Organizations"

Daniel H. Kim, The MIT Center for Organizational Learning, and Pegasus Communications

Keith Perry, McClellan Air Force Base

Kellie Wardman O'Reilly, Pegasus Communications

Carol Ann Zulauf, Suffolk University

and, finally, all the folks at Coca-Cola:

Michael Canning, Connie Carroad, Dianne Culhane, Teresa Hogan, Scott Moyer, Cheryl Oates, and Rodolfo Salgado

We hope you'll find your adventure into systems thinking stimulating and rewarding on both a personal and a professional level. We also invite you to contact us with any comments, questions, or suggestions about using this book or about systems thinking in general.

Bon voyage!

Virginia ("Prinny") Anderson   (VRAnderson@aol.com)
Lauren Johnson   (ljohnson@pegasuscom.com)

What Are Systems?

Welcome to the world of systems and systems thinking! You may be asking yourself, Why is it important to explore systems? One reason is that we live in and are influenced by systems all around us, from the natural environment to healthcare, education, government, and family and organizational life. Understanding how these systems work lets us function more effectively and proactively within them. The more we build our understanding of system behavior, the more we can anticipate that behavior and work *with* the system to shape the quality of our lives.

This section introduces you to the idea of systems and what makes them unique. In the learning activities at the end of the section, you will have the opportunity to identify some major systems in your own work life and to think about typical system behavior.

WHAT IS A SYSTEM?

A system is a group of interacting, interrelated, or interdependent components that form a complex and unified whole. A system's components can be physical objects that you can touch, such as the various parts that make up a car. The components can also be *intangible,* such as processes; relationships; company policies; information flows; interpersonal interactions; and internal states of mind such as feelings, values, and beliefs.

In an organizational setting, for example, the R&D group is a system made up of people, equipment, and processes that create new products to be manufactured by the production system and sold by the sales system. The components of the R&D group have to interact with one another to perform their function and thus are interdependent. In turn, the R&D group interacts and is interdependent with other systems within the company. A system such as the R&D group always has a specific purpose in relation to an even larger system—in this case, the entire organization (Figure 1.1, "Interdependent Systems Within Interdependent Systems").

Your body is another example. Within it, your circulatory system delivers oxygen, nutrients, hormones, and antibodies produced by other systems and carries waste to the excretory system. The circulatory system is made up of the heart, veins and arteries, blood, and a host of supporting elements. All of these components interact to carry out their purpose within the larger system—your entire body.

Both of these examples raise an intriguing point about systems: We can think of all systems as nodes embedded in a giant network in which everything is connected. For example, the company described above, with its interdependent R&D, production, and sales systems, is itself a large system that is interdependent with an even larger system—industry as a whole.

FIGURE 1.1

Interdependent Systems Within Interdependent Systems

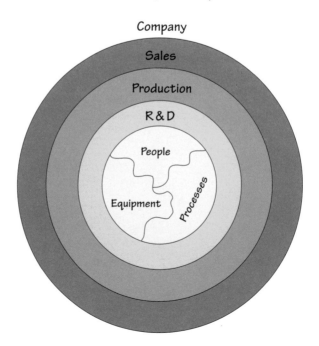

And industry is interdependent with an even larger system—the economy—and so on. The more we widen our view in this way, the more we see that everything—from the tiniest subatomic particle to the universe (and maybe beyond!)—is intertwined.

We can also distinguish between natural systems and human-made, nonliving systems. Natural systems—a living being's body, human societies, an ecosystem such as a prairie—have an enormous number and complexity of components and interactions among those components. They also have virtually an infinite number of connections to all the systems around them. Human-made systems—cars, for example—can also be quite complex, but these nonliving systems are not as intricately linked to systems around them. If a car breaks down, the impact of this event is not nearly as far-reaching as if a species were removed from a prairie ecosystem (although you may disagree if it's *your* car that breaks down!). Put another way, human-made nonliving systems are more self-contained than natural systems, which we can think of as more open in their connections to surrounding systems.

Defining Characteristics of Systems

Systems have several essential characteristics:

1. **A system's parts must all be present for the system to carry out its purpose optimally.**

If you can take components away from something without affecting its functioning and its relationships, then you have just a collection, not a system. For example, if you remove a cashew from a bowl of mixed nuts, you have fewer nuts, but you have not changed the nature of the collection of components. Therefore, a bowl of mixed nuts is not a system.

Similarly, if you can add components to a collection without affecting its functioning and relationships, it's still just a collection. So, if you add pistachios to your bowl of mixed nuts, you have more nuts and you have a different mix, but you still have just a collection of nuts.

However, if you assign new tasks to an R&D group or redefine the job descriptions of its staff, you will likely change the group's functioning and relationships—whether for the better or worse. The R&D group is not just an assortment of people, equipment, and processes; it is a system.

2. **A system's parts must be arranged in a specific way for the system to carry out its purpose.**

If the components of a collection can be combined in any random order, then they do not make up a system. For example, in a bowl of fruit, the oranges can go at the bottom, in the middle, or on the top without changing the essential nature of the collection of fruit.

However, in a system such as a company, imagine what would happen if the parts shifted around randomly—if, for instance, the accounting specialists suddenly decided to work on the production line, and the production specialists decided to write marketing copy. Of course, people do change jobs within their companies, but only after training and much transition time. Most companies function best when people are working in jobs that match their skills and experience, and when the staff is organized according to a specific plan.

3. Systems have specific purposes within larger systems.

All systems have a specific purpose in relationship to the larger system in which they're embedded, as we saw in the examples of the R&D department and human circulatory system above. Because each system has its own purpose, each is a discrete entity and has a kind of integrity that holds it together. In other words, you can't force two or more systems together and get a new, single, larger system. Nor can you subdivide a system and automatically end up with two smaller identical, functioning systems.

As the saying goes, if you divide an elephant in half, you don't end up with two smaller elephants. And if you put two small elephants together, you don't have a new, single, larger elephant (although some day you may end up with a new system—known as a herd!).

4. Systems maintain their stability through fluctuations and adjustments.

Left to themselves, systems seek to maintain their stability. Your organization does its best to maintain a designated profit margin just as most human bodies work to maintain a temperature of about 98.6 degrees Fahrenheit. If you examined your organization's revenues against expenditures every week or graphed your body temperature every five minutes, you would probably draw a wobbly, fluctuating line that nevertheless holds steady *overall*. Margins appear and disappear as a company pays its suppliers and collects checks from customers. Your body temperature rises and falls depending on your mood and your level of physical exertion. On average, however, your body temperature remains stable. And, with reasonable management and no cataclysmic change, your organization's margin also remains stable overall.

Systems achieve this stability through the interactions, feedback, and adjustments that continually circulate among the system parts, and between the system and its environment. Let's say a corporation receives an unusually large stack of suppliers' invoices (external stimuli) in the mail. The accounts payable department responds by paying the bills. As the checks go out the door, the accounting department, alarmed, compares revenue versus expenditures and gives feedback to management: Expenditures are up and revenues aren't covering them. Management then adjusts the system by reminding key customers to pay overdue invoices. Similarly, if you go for a run, your exertion warms your body. The sensation of heat is fed back to your sweat glands, which begin to work. Over time, sweating readjusts your temperature back to the norm.

5. Systems have feedback.

Feedback is the transmission and return of information. For example, imagine that you are steering your car into a curve. If you turn too sharply, you receive visual cues and internal sensations that inform you that you are turning too much. You then make adjustments to correct the degree of your turn. The most important feature of feedback is that it provides the catalyst for a change in behavior.

A system has feedback within itself. But because all systems are part of larger systems, a system also has feedback between itself and external systems. In some systems, the feedback and adjustment processes happen so quickly that it is relatively easy for an observer to follow. In other systems, it may take a long time before the feedback is returned, so an observer would have trouble identifying the action that prompted the feedback. For example, if you sunbathed a lot in your teens, you may develop skin problems after age 40—but because so much time passed between the two events, you may not recognize the connection between them.

Finally, feedback is not necessarily transmitted and returned through the same system component—or even through the same system. It may travel through several intervening components within the system first, or return from an external system, before finally arriving again at the component where it started.

For instance, imagine that the company you work for is suffering financially and decides to lay off 20 percent of the work force. That quarter, the layoff does indeed improve the looks of the financial bottom line. On this basis, the upper management might decide that layoffs are a reliable way to improve the financial picture.

However, let's say you survived the layoff; how would you describe your state of mind and that of your other remaining colleagues? Besides cutting costs, layoffs are also famous for damaging morale and driving people to "jump ship" in search of more secure waters. Eventually, as low morale persists, you and your colleagues might start coming to work late and leaving early, and caring less and less about the quality of your work. Productivity could drop. In addition, everyone who leaves—whether voluntarily or by being laid off—takes valuable skills and experience with them, so the overall capability of the work force goes down, further hurting productivity. Lowered productivity leads to expensive mistakes and lost sales from disgruntled customers. All this eats away even more at the company's revenue, tempting management to think about having even more layoffs to cut costs.

In this example, the feedback that made layoffs look like good policy was returned quickly—probably within one quarter. The feedback about the long-term costs of layoffs went through more steps and took a lot longer to return. Yet this information was essential for the management team to see the full impact of their decisions.

 ## EVENTS, PATTERNS, STRUCTURE

In reading all this information, you may be wondering what actually gives rise to systems. Systems are built on structures that leave evidence of their presence, like fingerprints or tire marks, even if you can't see them. But what is structure, exactly? The concept is difficult to describe. In simplest terms, structure is the overall way in which the system components are interrelated—the organization of a system. Because structure is defined by the *interrelationships* of a system's parts, and not the parts themselves, structure is invisible. (As we'll see later, however, there are ways to draw our *understanding* of a system's structure.)

Why is it important to understand a system's structure? Because it's system structure that gives rise to—that explains—all the events and trends that we see happening in the world around us.

Perhaps the best way to grasp the role of structure is to explore the Events / Patterns / Structure pyramid, shown in Figure 1.2.

Events

We live in an event-focused society (Figure 1.3, "The Tip of the Pyramid"). A fire breaks out in the neighborhood; a project misses a deadline; a machine breaks down. We tend to focus on events rather than think about their causes or how they fit into a larger pattern. This isn't surprising; in our evolutionary development as a species, this ability to respond to immediate events ensured our very survival.

But focusing on events is like wearing blinders: You can only *react* to each new event rather than anticipate and shape them. What's more, solutions designed at the event level tend to be short lived. Most important, they do nothing to alter the fundamental structure that caused that event. For example, if a building is burning, you would want local firefighters to react by putting out the fire. This is a necessary and essential action. However, if it is the *only* action ever taken, it is inadequate from a systems thinking perspective. Why? Because it has solved the immediate problem but hasn't changed the underlying structure that caused the fire, such as inadequate building codes, lack of sprinkler systems, and so on.

By uncovering the elusive systemic structure that drives events, you can begin identifying higher-leverage actions. The next step to comprehending systemic structure is to move from thinking at the event level to thinking at the pattern level.

Patterns

Whereas events are like a snapshot, a picture of a single moment in time, patterns let us understand reality at a deeper level (Figure 1.4, "Moving from Events to Patterns"). Patterns are trends, or changes in events over

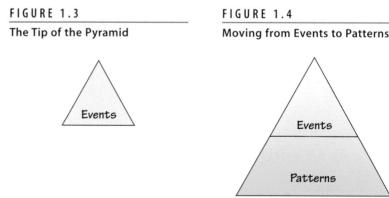

FIGURE 1.2

The Events / Patterns / Structure Pyramid

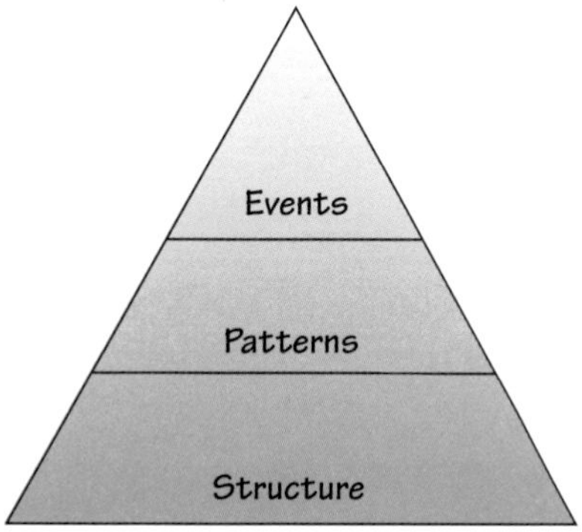

FIGURE 1.3

The Tip of the Pyramid

FIGURE 1.4

Moving from Events to Patterns

FIGURE 1.5

Graphs of Patterns

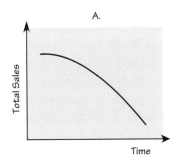 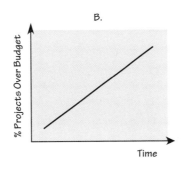 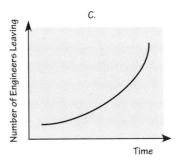

time. Whenever you see a pattern of events—for example, sales have been declining over the past few years, or two-thirds of the department's projects have gone over budget in the last year, or several senior engineers have left the company recently, most of them in the last six months—you're getting one step closer to grasping the systemic structure driving that pattern.

In each of the above examples, you could draw a simple graph to represent the trend (Figure 1.5, "Graphs of Patterns").

What is the advantage of thinking at the pattern level, as opposed to the event level? Detecting a pattern helps you put the most recent event in the context of other, similar events. The spotlight is then taken off the specific event, and you can focus on exploring how the series of events are related and begin thinking about what caused them. In the end, to anticipate events and ultimately change a pattern, you need to shift your thinking one more time: to the level of structure (Figure 1.6, "The Complete Pyramid").

Structure

To move to this deeper level of understanding, let's reconsider the above example of the senior engineers' exodus. You might begin digging for the structure behind this pattern by asking, "What's causing more and more senior engineers to leave?" In this case, suppose a change in corporate policy has cut both the budget and the number of administrative assistants for the engineering group. The engineers' workloads have ballooned, and they've begun grumbling more and more about their job pressure. Worse yet, as some of them leave, those left behind get even more upset as their workloads expand further. It's a vicious cycle that you might sketch as shown in Figure 1.7, "The Engineering Exodus," p. 8.

FIGURE 1.6

The Complete Pyramid

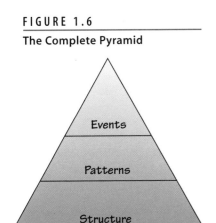

FIGURE 1.7

The Engineering Exodus

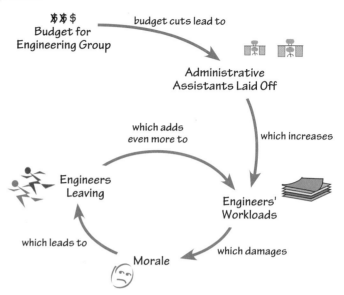

Whenever we ask questions like, "Why is this pattern happening?" or "What's causing these events?" we are probing at structure. Thinking at the structural level means thinking in terms of causal connections. *It is the structural level that holds the key to lasting, high-leverage change.* Let's return to our example about a house catching fire, to see how this works. To fight fires at the event level, you would simply *react* to quell the fire as soon as possible after it broke out. You would probably then repair any smoke and water damage, and put the incident out of your mind.

A NOTE ABOUT DIAGRAMMING SYSTEMS

As we saw earlier, all systems are part of larger systems, so it's impossible to capture any system in its entirety on paper. Nevertheless, there are ways of depicting *parts* of systems in a diagram, in order to glimpse how a system works and how you might alter its behavior. One such way is to create a causal loop diagram, or CLD. (Figure 1.7 is an example of this kind of drawing.) These diagrams provide a starting place for discussing and thinking about problematic events or patterns, and for opening the door to addressing problems differently. In particular, they help you gain insight into systemic structures, and they identify ways you might change the system's behavior. After all, it is changes made at the system level, rather than at the pattern or event level, that often prove to be the most long-lasting and self-sustaining.

It's important to remember, however, that graphic representations of systems are just that: representa-tions. As you will see later in this book, there is no one right way to draw a causal loop diagram or even to describe an entire system. Any diagram that you draw reflects your own assumptions about the system, and is limited to what you define as the most pertinent part of the system you're studying. This is why working in groups is so beneficial—you gain insights from the multiple perspectives.

Later in this book, you'll have the opportunity to practice drawing CLDs. As with all the activities in this book, we encourage you to work as a group whenever possible in creating CLDs. A causal loop diagram generated by a group is especially valuable because it reveals the interplay of each group member's perspective on the system in question. The process of constructing the drawing encourages group members to share their assumptions and under-standings about the issue at hand. The more this kind of sharing happens, the more insights get sparked.

How would you fight fires at the pattern level? You would begin anticipating *where* other fires are most likely to occur. You may notice that certain neighborhoods seem to suffer more fires than others. You might locate more fire stations in those areas, and staff them based on past patterns of usage. By doing these things, you would be able to fight fires more effectively by *adapting* to the *patterns* you have observed.

However, your actions haven't done anything to reduce the actual occurrence of fires. To address the problem at this level, you need to think about the structure that gives rise to the pattern of fires. At the systemic structure level, you would ask questions like, "Are smoke detectors being used? What kinds of building materials are least flammable? What safety features reduce fatalities?" Actions that you take at this deep level can actually cut down the number and severity of fires. Establishing fire codes with requirements such as automatic sprinkler systems, fire-proof materials, fire walls, and fire alarm systems saves lives by preventing or containing fires.

Here's where the real power of structural-level thinking comes in: Actions taken at this level are *creative*, because they help you to shape a different future, the future that you want. Does this mean that high-leverage actions can be found only at the structural level? No—leverage is a relative concept, not an absolute. Our ability to influence the *future* increases as we move from event-level to pattern-level to structural-level thinking, but sometimes the best action we can take must remain focused on the present, at the event level—for example, when a building is aflame, the highest leverage action in the moment is to react by putting out the fire. Any other action would be downright inappropriate. But, if that's all we did, the actions would be considered low leverage from a long-term perspective. The art of thinking at the systemic structure level comes with knowing when to address a problem at the event, pattern, or structural level, and when to use an approach that combines the three.

Figure 1.8, "Levels of Understanding," depicts the richness of these three levels of understanding.

FIGURE 1.8

Levels of Understanding

	Action Mode	Time Orientation	Way of Perceiving	Questions You Would Ask
Events	React!	Present	Witness event	"What's the fastest way to react to this event now?"
Patterns	Adapt!		Measure or track patterns of events	"What kinds of trends or patterns of events seem to be recurring?"
Structure	Create Change!	Future	Causal loop diagrams and other systems thinking tools	"What structures are in place that are causing these patterns?"

These activities can be done either by yourself or with a group. For self-study, you might consider starting a systems thinking journal, and doing the following learning activities over the course of a week or two. However you approach them, take plenty of time to think about each activity. Be honest, too. No one else will see your notes or your journal!

If you are helping a group to do the activities, have them read and think about the material ahead of the meeting time. Then go over it when you meet, answering questions and looking for additional examples to help illustrate key points.

ACTIVITY 1   IDENTIFYING SYSTEMS

Purpose: To identify systems and their components
To recognize interrelated systems

Outcome: Recognition of systems within your organization, their interrelationships, and their purposes
Insights about intangible and possibly powerful components of systems that affect what happens in your organization.

Instructions: Identify three systems in your organization, including at least one that includes some important but intangible components.

Example ➤   Informal information system

Purpose: To supplement the "official" information system so people feel they can make more informed decisions. The informal system may also help defuse tension by offering an avenue for chatting or gossiping.

Components: People, the electronic mail system

Intangible components: Information or "gossip," time to communicate, motivation to share information

Larger system: The overall information system, which also has a formal communication system

1. List your three systems below:

2. In the spaces provided below, fill in the following information:
    - The name of your system
    - The purpose that your system fulfills within the larger system
    - The components that make up your system
    - The intangible components of your system
    - The larger system of which your system is a part

Your First System:
Purpose:
Components:
Intangible components:
Larger system:

Your Second System:
Purpose:
Components:

Intangible components:

Larger system:

Your Third System:

Purpose:

Components:

Intangible components:

Larger system:

ACTIVITY 2   **REFLECTING ON SYSTEM CHARACTERISTICS**

Purpose: To think about the behavior of the systems you see around you
           To identify patterns of behavior over time, and think about what causes that behavior

Outcome: Simple graphs of observed behavior patterns
           Awareness of the forces that drive behavior patterns

Instructions: Write your answers to the following questions in the space provided.

QUESTIONS

1. Identify a chronic problem or ongoing issue that you wrestle with at work (for example, "Sales do well for a while, then drop, then pick up again"; or "Every year more and more people get laid off").

2. Try drawing a graph of what seems to be happening. (Tip: Ask yourself, Is the pattern going up? Going down? Oscillating over time? Going up or down and then leveling off?)

3. Looking at your graph, what do you see?

4. Is there any way that actions taken to solve the problem might actually be making things worse? If so, how?

ACTIVITY 3   UNDERSTANDING SYSTEM FEEDBACK

Purpose: To begin using simple diagrams to show how components of a system are related and how feedback is returned through the system

Outcome: Diagrams of the three systems you identified in Activity 1

Instructions: For each system you identified in Activity 1, draw a map or diagram of how the parts are related and how one part gives feedback to another. You may find there are many or only a few feedback connections. You may also find that some feedback travels through multiple steps before arriving back at its original source.

Example ➤   Figure 1.9, "The Informal Information System," shows the system discussed in Activity 1.

FIGURE 1.9

The Informal Information System

In this diagram, the dotted arrows represent feedback. The number of people who want to share information, the amount of information to share, and the amount of time available are all related to the level of information sharing. The amount of information sharing or its value to people can determine how much interest everyone has in using the electronic mail system to share information. This level of interest then influences how many people share information and how much time they spend doing it.

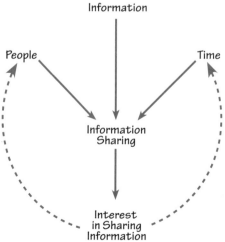

Now try diagramming your three systems:

Your First System:

Your Second System:

Your Third System:

Finally, try trading diagrams with someone else. Take turns explaining the diagrams and your understanding of the systems you drew.

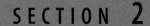

What Is Systems Thinking?

There are many ways of looking at systems thinking. It offers not only a set of tools, but also a framework for looking at issues as systemic wholes. For some people, it can even become a way of life! Systems thinking is a language, too, that offers a way to communicate about dynamic complexities and interdependencies. Most Western languages are linear—their basic sentence construction, noun-verb-noun, encourages a worldview of "*x* causes *y*." Because of this, we tend to focus on linear causal relationships rather than circular or mutually causative ones. Yet many of the most vexing problems confronting managers and corporations today are caused by a web of interconnected, circular relationships. To enhance our understanding and communication of such problems, we need a language and a set of tools better suited to the task. This is where systems thinking comes in.

In this section, we examine the foundational principles of systems thinking in more detail, and explore the special qualities of systems thinking as a language. The learning activities at the end of the section will let you begin actually practicing systems thinking.

THE PRINCIPLES OF SYSTEMS THINKING

In general, systems thinking is characterized by these principles:

- thinking of the "big picture"
- balancing short-term and long-term perspectives
- recognizing the dynamic, complex, and interdependent nature of systems
- taking into account both measurable and nonmeasurable factors
- remembering that we are all part of the systems in which we function, and that we each influence those systems even as we are being influenced by them

The "Big Picture"

During stressful times, we tend to focus on the immediate, most pressing problem. With this narrow focus, we can perceive only the *effects* of changes elsewhere in the system. One of the disciplines of systems thinking, however, involves being able to step back from that immediate focus and look at the bigger picture. As you know, whatever problem you're involved in right now is part of a larger system. To discover the source of a problem, you have to widen your focus to include that bigger system. With this wider perspective, you're more likely to find a more effective solution.

For example, imagine that you work for a regional appliance distributor that is experiencing growing delays in providing timely service to its customers. As a manager in this company, you might be tempted to focus first on the service technicians, perhaps on the service order-taking system, or even on service-order dispatching. But take a step back. What if you knew that about six weeks before the service crisis started, the sales group had implemented an incentive program that included free set-up and installation of new appliances? To sell more appliances, the sales force encouraged customers to make service appointments as soon as their delivery dates were set. Then a service person could come out to install the appliance and make all the electrical and water connections. However, because the customer service department had not been informed of the incentive program, it had no opportunity to add staff to handle the installation incentive. With this wider view, you might conclude that the delays in providing customer service do not come from the customer service department, and you might choose a different path to solving the problem.

Long Term, Short Term

How often does your organization expect to see results of its activities? In a year? A quarter? A week? In addition to checking the health of the company at these kinds of intervals, some businesses also make major strategic changes—such as cost-cutting campaigns, layoffs, new hiring, production increases—every time they check how the business is doing.

Yet systems thinking shows that behavior that leads to short-term success or that is prompted by short-term assessments can actually *hurt* long-term success. However, the point is not that the long-term view is "better" than the short-term view. After all, if a little boy runs out into traffic, grabbing him by the arm at risk of injuring his shoulder or startling him makes far more sense than moving slowly or speaking softly while a bus speeds down the street. In thinking about any decision, the best approach is to strike a balance, to consider short-term *and* long-term options and to look for the course of action that encompasses both. At the very least, try making your decisions by first thinking through their likely ramifications—both short term *and* long term.

Here's an example: As a business grows, it may use consultants to handle its human resource and training functions. In the short run, this sensible business decision can bring in a high level of professional expertise. If the company decides that consulting help is getting too expensive, however, it will eventually move to develop its own in-house HR and training department. To balance the long and the short term, the company could phase in internal expertise at certain levels of revenue, sales volume, or staffing, and overlap internal and outside resources while the new staff members get their bearings.

Whether you're focusing on the long term or the short term, the key is to be aware of all the potential impacts of whichever strategy you choose.

Dynamic, Complex, and Interdependent

When you look at the world systemically, it becomes clear that everything is dynamic, complex, and interdependent. Put another way: Things change all the time, life is messy, and everything is connected.

We may know all this. However, when we're struggling with an overwhelming problem or an uncertain future, we tend to want to simplify things, create order, and work with one problem at a time. Systems thinking doesn't advocate abandoning that approach altogether; instead, it reminds us that simplification, structure, and linear thinking have their limits, and can generate as many problems as they solve. The main point is that we need to be aware of all the system's relationships—both within it and external to it.

Measurable vs. Nonmeasurable Data

Some organizations value quantitative (measurable) over qualitative (nonmeasurable) data. Others are just the opposite. Systems thinking encourages the use of both kinds of data, from measurable information such as sales figures and costs to harder-to-quantify information like morale and customer attitudes. Neither kind of data is better; both are important.

Systems thinking also alerts us to our tendency to "see" only what we measure. If we focus our measuring on morale, working relationships, and teamwork, we might miss the important signals that only statistics can show us. On the other hand, if we stay riveted on "the numbers," on how many "widgets" go out the door, we could overlook an important, escalating conflict between the purchasing and the production departments.

We Are Part of the System

One of the more challenging systems thinking principles says that we usually contribute to our own problems. When we look at the big picture, over the long term, we often find that we've played some role in the problems facing us.

Unintended consequences. Sometimes the connection is simple—the problem plaguing us today is an unintended consequence of a solution we implemented yesterday. For example, to control costs, a bank manager decides to limit the number of tellers on Thursday evenings and Saturday mornings. Eventually, the manager notices that—surprise!—other banks seem to be getting all the customers who rely on having access to the bank during evenings and weekends.

Assumptions. Sometimes our assumptions are what get us into trouble. Imagine, for example, that you're the manager of Frank's Steak House, a restaurant that specializes in affordable family dining. You've noticed that business at the restaurant has flagged a bit for two or three months in a row. You conclude that this is an enduring trend, because you've read essays in the newspapers about a possible resurgence in the health of the national economy. People are feeling freer to dine at more expensive restaurants, you decide. To prepare Frank's to weather the new trend, you lay people off. However, demand bounces back a few months later, and you're forced to scramble to bring workers back. Some of these workers are rehired at higher pay than before, some on overtime. These kinds of assumptions about how the world works (also known as mental models) are powerful drivers of the decisions we make.

Values and beliefs. Deeply held values and beliefs can lock us into counterproductive ways of making decisions. The Cold War is a perfect example: As long as the U.S. and the former U.S.S.R. each firmly believed that the other was intent on annihilating its ideological enemy, the arms race was inevitable. Even worse, the longer the conflict continued, the harder it was to call it off. Both nations were highly invested in justifying their "saber-rattling" in the past, present, and future. In this case, too, mental models played a major role.

 ## SYSTEMS THINKING AS A SPECIAL LANGUAGE

As a language, systems thinking has unique qualities that make it a valuable tool for discussing complex systemic issues:

- It emphasizes looking at wholes rather than parts, and stresses the role of interconnections. Most important, as we saw earlier, it recognizes that we are part of the systems in which we function, and that we therefore contribute to how those systems behave.

- It is a circular rather than linear language. In other words, it focuses on "closed interdependencies," where *x* influences *y*, *y* influences *z*, and *z* comes back around to influence *x*.

- It has a precise set of rules that reduce the ambiguities and miscommunications that can crop up when we talk with others about complex issues.

- It offers visual tools, such as causal loop diagrams and behavior over time graphs. These diagrams are rich in implications and insights. They also facilitate learning because they are graphic and therefore are often easier to remember than written words. Finally, they defuse the defensiveness that can arise in a discussion, because they emphasize the dynamics of a problem, not individual blame.

- It opens a window on our mental models, translating our individual perceptions into explicit pictures that can reveal subtle yet meaningful differences in viewpoints.

To sum up, the language of systems thinking offers a whole different way to communicate about the way we see the world, and to work together more productively on understanding and solving complex problems.

LEARNING ACTIVITIES

In this section, each learning activity focuses on one or two systems principles. As with the Section 1 learning activities, the exercises here can be done either by yourself or with a group. If you are working with a group, focus on the activities that highlight principles you consider the most valuable for your organization. Keep in mind that some of the activities are active exercises; some are meant for individual reflection and group discussion—try to use a mix.

Finally, remember that there is no one right response to the exercises. The idea is to use your imagination, and to have some fun!

Activity 1   STRETCHING THE TIMELINE

Purpose: To think in "big picture" terms
To consider both short-term and long-term perspectives on a problem
To practice seeing patterns and trends in a problem
To identify the roots of a current problem

Outcome: A timeline showing the history of a current problem
Insights about recurring patterns in an organization

Number: Minimum 1; maximum about 15

Equipment: For self-study: A white board or a couple of flip-chart pages and colored markers
For a group: Colored yarn, pushpins or tape, several pairs of scissors, and a large wall or floor space OR long sheets of flip-chart paper, three or more colored markers for each person

Space: For a group, enough wall or floor space so that pairs or trios of people have at least six feet to themselves

Steps

1. Lay your flip-chart paper on its side, horizontally, and position yourself near the far right end of the page. (You might even want to tape two flip-chart pages together horizontally, to give yourself lots of writing space.) If you are working at a white board, adapt the directions accordingly.

2. Identify a current problem or issue facing your immediate work group or department. If nothing comes to mind, use a problem within your family or community. Choose a moderate-size issue with which you have direct personal experience.

3. Make a mark on the paper to symbolize the present, and name the issue in one or two words. For example:

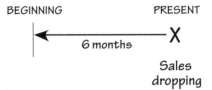

4. To the best of your knowledge, when did this problem start? Pick a distance to the left across the paper that represents the amount of time you think has elapsed since the problem began. Mark the beginning point with your marker. Draw a line between the beginning point and the present. Write in the time span.

5. Now project yourself back in time to the "Beginning" point. To the best of your knowledge, what was happening around that time to cause the beginning of the problem? Write your answer as a brief phrase, as shown in the example below.

6. With a new marker color, extend your timeline even farther back in time, as shown below. Add a time span between "Earlier beginning" and "Beginning."

7. Now project yourself back once more, to "Earlier beginning." Think of what was happening at *that* point that led to the problem you wrote under "Beginning." Name it and mark it on the timeline in a third color.

8. Continue the process one more time, by adding "Earliest beginning" as shown below. Add what was happening at *that* time, and fill in a time span between "Earliest beginning" and "Earlier beginning."

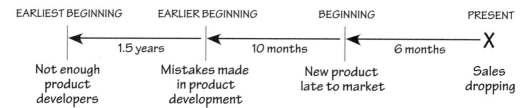

EARLIEST BEGINNING	EARLIER BEGINNING	BEGINNING	PRESENT
1.5 years	10 months	6 months	X
Not enough product developers	Mistakes made in product development	New product late to market	Sales dropping

9. Now imagine yourself present at any of the three beginning points you have identified. Is there anything else going on at that time that resembles the original problem you chose? Or is there another kind of problem that alternates with yours? Is there another problem going on in parallel to yours right now? (In the timeline boxes above, for example, maybe there was a hiring difficulty going on at the "Earliest beginning" stage of the problem.) If you can identify a parallel problem, add it to your timeline as a parallel line, using a distinctive color.

10. Now try one more step with your timeline. Instead of projecting backward in time, project forward. Given the sequence of events you've drawn, what do you expect to happen in the future if nothing is done about the problem? Add another piece of flip-chart paper if necessary, and extend your timeline to the right. Using a new marker color, add your thoughts about the future to this new part of your timeline.

QUESTIONS

1. What was it like to create a visual image of the time and events surrounding the current problem you identified? Any surprises? Any insights? Write your thoughts below. If you worked on this activity with a group, discuss your insights together.

2. What did you learn?

3. If you were able to identify parallel problem timelines in Step 9, what did you learn?

4. If you were *not* able to trace back to earlier beginnings, what did you learn?

5. What helps you to see the "big picture" of your problem?

6. What obscures it?

<table>
<tr><td>*ACTIVITY 2*</td><td>**THE SHAPE OF THE PROBLEM**</td></tr>
</table>

Purpose:	To explore the connections and interdependencies among the components of a problem
	To discover the intangible aspects of a problem
	To practice widening your view of a problem
	To see the complexity within a problem
Outcomes:	A map of the connections and interdependencies of a problem
	Insights about the structure of the problem
Number:	Minimum 1; maximum 20
Equipment:	Flip-chart paper and three or four colored markers per person
Space:	For groups, enough wall, floor, or table space for everyone to lay out a flip-chart page and draw

STEPS

1. Identify a problem or an issue currently facing you or your immediate work group. (Your work group might be your department, division, unit, and so forth.) In the center of your flip-chart paper, draw a circle and write in the name of your group. Write one or two words to identify the issue you chose, as shown in the example in Figure 2.1, "The Center Circle."

FIGURE 2.1

The Center Circle

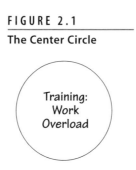

2. Who else outside your group is directly involved in or affected by this issue? Write your answers within their own circles in a ring around your central issue. Connect each outer circle with the middle circle, similar to Figure 2.2, "The Circle Expands."

FIGURE 2.2

The Circle Expands

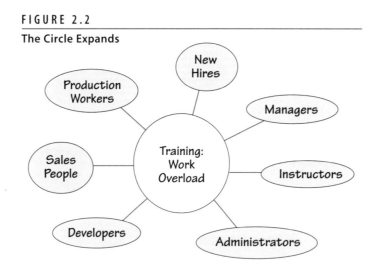

3. Who is touched by each of the individuals or groups you identified in Step 2? Who is indirectly connected to your issue or problem? Don't forget families, friends, and other groups that can be affected when people are stressed, working overtime, excited by their successes, or receiving bonuses. Draw these people or groups into the picture and connect them to the appropriate circles, as in Figure 2.3, "Even More Connections."

FIGURE 2.3

Even More Connections

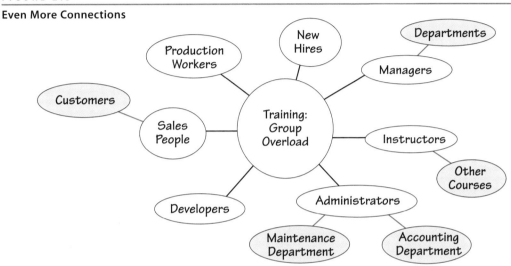

4. If there are any other connections beyond what you've already drawn, map them in. The connections are not limited to human beings. They can include items such as "Revenues" or "Other Groups' Work." Your map can have as many circles or layers as make sense to you, as shown in Figure 2.4, "The Final Picture."

FIGURE 2.4

The Final Picture

QUESTIONS

1. In your diagram, what happens to the *outer* circles when things are going well in the center circle? When they're not going well? Examples?

2. What happens to the *center* circle when things are going well in the other circles? When they're not going well? Examples?

3. Looking at the interconnections, can you see any ways in which something you do in the center circle causes a change in a connected circle that then comes back and affects the center circle? Examples?

4. Did you find it difficult to add many circles to your original circle? If so, what are some possible reasons for this difficulty?

5. If you worked on your map with others, discuss your insights together. If you worked with others, but each of you made your own map, exchange your maps and share the insights about the maps.

ACTIVITY 3   **IS TIME ON YOUR SIDE?**

Purpose: To think about how your organization sets goals, and how frequently it measures results
To explore the impact of the time cycles involved in setting goals and measuring results

Outcome: Insights about how time frames influence what we pay attention to and what we accomplish

Number: Minimum 1; maximum as many as desired

Equipment: Flip chart and markers (optional)

Instructions: Reflect on the following questions, and discuss them with others if possible.

QUESTIONS

1. What is your organization's stated goal or mission? What is it trying to achieve? (State the mission as simply as possible. "Organization" can refer to your immediate work group, your division or department, or the overall organization.)

2. Is there a desired time frame for achieving the goal or mission? If so, what is it?

3. What results does the organization measure or pay attention to? (Examples: sales volume, revenue, meals served, passenger miles, return on assets, return to shareholders)

4. How often does the organization measure those results? (Examples: sales volume per quarter, meals served per week, passenger miles per vehicle, return on assets per year)

5. What goals does the organization have regarding what it measures? (Example: 2 percent sales growth per quarter)

6. How long does the organization take to produce, create, or deliver what is measured? (Examples: selling groceries takes 5–10 minutes; selling a car takes 1 hour–2 weeks; selling a house takes 1 day–1 year; selling a large management-information system takes 6–18 months)

7. What do you notice about the time frames for your organization's mission; for its target results; for its measurements; and for production, service, or delivery?

8. What do you think are the effects of those time horizons?

ACTIVITY 4   **FROM SHORT TERM TO LONG TERM**

    Purpose: To discover which aspects of your work are short term and which are long term

    Outcome: Timelines of short-term and long-term events or outcomes

    Number: Minimum 1; maximum as many as desired

    Equipment: Flip-chart paper, tape, and markers OR lined paper and pens or pencils

Round 1: Your Organization—Short Term or Long Term?

    Instructions: Write your answers to each question in the accompanying box.

QUESTIONS

1. What is the shortest-term discrete product, service, or other deliverable from your organization? How long does it take to produce or deliver it?

    Examples ➤ A soft drink / 90 seconds to fill and serve
                 A tank of gas / 5 minutes to fill
                 A consultation / 1 hour
                 An express package / 15 hours from pick-up to delivery
                 A house / 4 months from ground-breaking to finished siding

Shortest-term deliverable: *How long?*

```
┌────────────────────────┐    ┌────────────────────────┐
│                        │    │                        │
│                        │    │                        │
│                        │    │                        │
│                        │    │                        │
└────────────────────────┘    └────────────────────────┘
```

2. What is the longest-term product, service, or other deliverable from your organization? How long does it take to produce or deliver it?

   Examples ➤ A bridge / 3 years
   A communication system / 18 months
   A new management competency / 12 months

Longest-term deliverable: *How long?*

```
┌────────────────────────┐    ┌────────────────────────┐
│                        │    │                        │
│                        │    │                        │
│                        │    │                        │
│                        │    │                        │
└────────────────────────┘    └────────────────────────┘
```

3. What, if anything, falls into a middle-term length of time? How long does it take to produce or deliver it?

   Middle-term deliverable: *How long?*

```
┌────────────────────────┐    ┌────────────────────────┐
│                        │    │                        │
│                        │    │                        │
│                        │    │                        │
│                        │    │                        │
└────────────────────────┘    └────────────────────────┘
```

4. How much of your routine work is spent on the short-term end of the spectrum? At the long-term end?

   % Short-term: *% Long-term:*

```
┌────────────────────────┐    ┌────────────────────────┐
│                        │    │                        │
│                        │    │                        │
│                        │    │                        │
│                        │    │                        │
└────────────────────────┘    └────────────────────────┘
```

5. Looking at your answers, how would you define "short term" and "long term" in your organization?

6. How do you think your organization's sense of short and long term compares to other organizations'? How does this sense differ? How is it similar?

7. Where is the emphasis in your organization—long term or short term? Why? What drives that focus?

Round 2: You—Short Term or Long Term?

Instructions: In the space provided, jot down your responses to the following questions.

1. What do you want to accomplish today?

2. This week?

3. This month?

4. This year?

5. Within five years?

6. Within 10 years?

7. By the time you're very old?

8. Looking at your answers to the above questions, how would you define "short term" and "long term" in your own life? At what point is short term differentiated from long term?

9. How do you think your sense of short and long term compares to your organization's? How does this sense differ? How is it similar?

10. What do you emphasize in your own life—long term or short term? Why? What drives that focus?

11. When you made your list, was there a point at which your vision of what you want to accomplish shifted? If so, where, and how?

12. Do you think this kind of shift happens within your organization, too? If so, at what point in the timeline?

ACTIVITY 5   **IN THE MIDST OF A PROBLEM**

Purpose: To gain familiarity with the concepts of interconnectedness or interdependency
           To recognize the human tendency to assign blame

Outcome: Insights about our role in the problems we experience

Number: Minimum 1; maximum as many as desired

Instructions: Write your answers to the following questions in the space provided.

QUESTIONS

1. Briefly describe a situation in which you knew that an individual or group having a problem was contributing to the problem, but wasn't aware of their contribution.

Example ➤

I used to work with someone, Valerie, who swore a lot at the office—really rough language. One day, she came in upset because her eight-year-old daughter, Nina, had been sent home from school for swearing. Valerie couldn't understand where Nina picked up this behavior! It was so obvious to the rest of us, but she just couldn't see it.

2. Now describe a situation in which you or your work group turned out to be contributing to your own problem.

    Example ➤

    I was experiencing deteriorating communications with a senior project team leader, Alan. I tried to clarify the relationship—I left him voice mails and got no response. I sent him memos and heard nothing back. Projects came up that I was perfect for, but Alan didn't include me. I was *furious* with him. When I finally managed to meet with him, I discovered that he was communicating less with me because he felt confident about our relationship and had other problems to take care of. My deluge of voice mails and memos made him think I was under a lot of stress, so when it came time to staff demanding projects, he decided to give me a break and leave me off. At the same time, though, Alan was beginning to wonder if I was becoming unreliable.

3. Consider a persistent, recurrent, or chronic problem you are experiencing now. Tell or record the story of the problem very briefly:

4. Now ask yourself:

    A. Is there any way you or your group may be causing or contributing to the problem? If so, how?

B. Is there anything you did in the past that has generated an unintended consequence? If so, what?

C. What might happen if you were to focus on the short-term aspects of the problem and ignore the longer term?

D. Sometimes feedback comes to you slowly or in roundabout ways. What, if any, aspect of the problem might stem from delayed or indirect feedback?

5. Do you now have any new insights into your problem? If so, what are they?

6. What, if any, difference does it make to see the part you are playing in a problem?

Uncovering Systemic Structures:

Drawing Behavior Over Time Graphs

I n sections 1 and 2, we introduced the idea that systemic structures generate patterns of behavior and are therefore at the root of many of our problems. In this section, we explore several steps for uncovering these structures:

1. Formulating the problem

2. Identifying the key variables in the situation; in other words, the main actors in the systemic structure

3. Graphing the behavior of those variables over time

Section 4 then takes you through the next step in identifying systemic structure: building causal loop diagrams.

As you read Section 3, remember that thinking systemically is an experimental process involving trial and error. The guidelines and the examples in this book may look orderly and straightforward, but applying systems thinking in real life is often messy and leads to lots of twists and turns. Thinking systemically always involves an iterative process of formulating problems with care, creating hypotheses to explain what is going on, tracking and revising the reasoning behind your explanations, testing possible solutions to problems, and reformulating the problem based on new understandings.

FORMULATING A PROBLEM

Let's say you've just finished a course on systems thinking and have identified a problem you want to address. Could you apply systems thinking tools to figure it out? Of course! All problems have systemic origins; the key is to choose one that is *appropriate* and *significant* to you. Here are some tips:

Guidelines for Identifying Systemic Problems

1. The problem is chronic and recurring.

2. The problem has been around long enough to have a history.

3. You or someone else may have tried to solve this problem, but your attempts either did not work at all or stopped working after a while.

4. You haven't been able to identify an obvious reason for the pattern of behavior over time.

5. The pattern of the problem's behavior over time shows one of the classic shapes in Figure 3.1, "Patterns of Problem Behavior."

Another reason for doing a systems thinking analysis is that the problem is important to you or to your organization, and is worth spending time and

FIGURE 3.1

Patterns of Problem Behavior

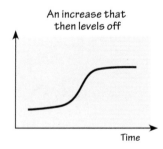

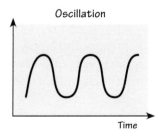
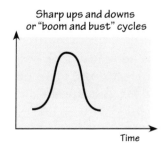
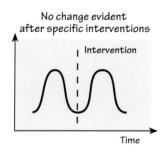

effort on solving it. Maybe the problem is currently under discussion, or you have a hunch that an old problem is about to strike again.

Here are some examples of problems that show typical systemic behavior:

- We've been having trouble getting our refrigerators assembled fast enough to fulfill customer orders. So, we reconfigured the flow of materials on the manufacturing floor to try to improve the assembly process. However, after we made this change, the assembly process actually took more time than before. Somehow, the change seems to have made everything worse.

- We introduced a line of high-grade investment portfolio products two years ago, but our agents continue to sell the older products. We've tried changing the incentive schemes, and we've put out stacks of marketing and information materials, but nothing seems to motivate agents to focus on the newer line.

- Every six months we go through another round of cost-cutting campaigns, from laying off workers to lengthening maintenance intervals to simplifying marketing. Costs go down for a while but then start rising again.

Guidelines for Formulating the Problem

Once you've targeted a problem for a systemic approach, work on developing a clear, succinct statement of the problem. This is often the toughest part of systems thinking, but it's very worthwhile. The more clearly and specifically you can state your issue, the more focused your systemic analysis will be. Be sure to brainstorm with other people who can contribute their views of the issue as well as their knowledge of its history. If necessary, proceed with two or three formulations of the problem and learn from the different views.

Don't get discouraged! It's natural—even beneficial—for this stage of the process to take a while. You and your group will generate the most insights into the problem by taking the time to ask lots of probing questions, share your perspectives on the issue, and revise your problem statement several—sometimes many—times.

Here are some examples of problem statements:

- In our blood lab, errors in sample analyses have doubled over the last eight months.

- Customer-service problems have increased 25 percent over the last year.

- Before our last two training conferences, we failed to return one-third of the registration confirmations to our customers on time.

Problem statements often include the following components (though they don't have to):

- the behavior (example: customer-service problems)

- a description of the behavior over time (example: the problems have increased)

- a measurement of how the behavior has changed over time
(example: the problems have increased 25 percent)
- the time frame of the behavior
(example: the problems have increased 25 percent in the last year)

 ## IDENTIFYING VARIABLES

Once you formulate the problem, it's time to identify its key variables. (Remember, variables are the components of the problem whose value can vary over time; that is, go up or down.) To begin this process, tell the story of the problem briefly. Telling the story means building on your problem statement—fleshing out some of the details so that you have a fuller picture of the issue and the variables involved.

Example ➤   ### The Case of A-to-Z

> At A-to-Z, a semiconductor company, we've been puzzling over a series of events that occurred in our most recent quarter. We posted record sales for the quarter, with the majority of our sales force meeting or exceeding sales quotas. All products scheduled for release were launched, with additional products ready for early release in the next quarter. At the same time, however, our profits actually declined for the first time in our company's history, as overhead costs as a percentage of sales reached an all-time high.

What are the pertinent variables in A-to-Z's story? Here's the list that A-to-Z's managers identified after some discussion:

Sales
New product releases
Profits
Sales force

Guidelines for Identifying Variables

Deciding *which* variables to work with is as important as clearly defining the problem, because your choice of variables shapes the rest of your analysis. As with formulating the problem, identifying variables is usually an iterative process. You might start off listing several, and then after much discussion and thought, decide to delete some or add new ones, or go back and rework your story of the problem.

Here are some guidelines for starting out:

- List all the variables that could reasonably be included, both quantitative and qualitative. For example, your list could include measurable variables such as "Sales" and "Size of sales force," as well as hard-to-measure variables like "Morale" and "Commitment to company goals." Again, do this as a group if possible, to get input from a wide variety of viewpoints. The idea is to start off with a big list of variables.

- Narrow your list down to the most important variables. You can do this by combining some variables because they represent roughly the same kind of information—for example, "Morale" and "Job satisfaction." You can also remove some variables from your list because you and your group decide that they're not as firmly linked to the central problem as other variables. To determine the most relevant variables, identify which variables seem to play the most prominent roles in the central issue you described in your problem statement. These variables will likely have a relationship to each other that you either can describe or want to explore. For example, in the case of A-to-Z's declining profits, the variables "Profits," "Sales," and "New product releases" are related to what the company defined as the central problem—and they have a significant dynamic relationship to each other.

As with every stage of the systems thinking process, you may decide to go back and revise your problem statement if your list of variables gives you new ideas about the nature or scope of your problem.

Guidelines for Naming Variables

After choosing your variables, it's time to refine their names precisely. The guidelines below will be especially important when you move on to creating causal loop diagrams in Section 4.

- Use nouns or noun phrases, *not* verbs or verb phrases.

Example ➤ New products in the pipeline
Revenues
Experience level of engineers

Not:
Developing new products
Being profitable
Sell
Produce

- A well-named variable fits into phrases such as "the level of," "the amount of," "the number of," "the size of".

Example ➤ The number of new products in the pipeline
The amount of revenues
The experience level of the engineers
The size of the profit margin

- Use a neutral or positive term whenever possible to name a variable.

Example ➤ "Job satisfaction" rather than "Job dissatisfaction"
"Morale" rather than "Bad feelings"

That way, you'll be able to describe the way the variable changes ("increases," "decreases," "improves," "worsens") without introducing confusing double-negatives. For example, the phrase "Job satisfaction declined" is much easier to grasp conceptually than "Job dissatisfaction declined."

- Keep in mind that variables can be concrete entities such as memory chips, buildings, or production workers, as well as intangibles such as morale, job satisfaction, or alignment with company values.

DRAWING BEHAVIOR OVER TIME GRAPHS

Once you've formulated your problem and chosen and named its variables, the next step is to graph the variables' behavior over time. You can then use the graphs to hypothesize about the variables' interrelationships, and to generate additional graphs that lead to deeper understanding of the problem. To draw behavior over time graphs (BOTs), we recommend the following three steps:

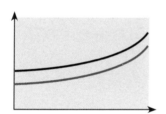

1. Select a time horizon.

2. Sketch the graph.

3. Build theories about how the graph's variables are interrelated.

Selecting a Time Horizon

Choosing a time horizon for your graph is an important decision, because the time horizon affects the amount and kind of information your graph will ultimately depict. Here are some guidelines:

1. Pick the variable with the longest time cycle—for example, new product development, production, or sales cycle—and extend the time horizon to cover three or more of those cycles, if possible.

2. Try to work with a minimum of two years, and experiment with five or more years. If you need to work with a much shorter or much longer time horizon, check your reasoning with your colleagues.

3. Sketch a timeline as shown in the example below, in which:

   Now is the present moment in which you are analyzing the problem.

   Earlier is the point earlier in time, two to five years ago, where you will begin tracing the behavior of the variables.

   Earliest is a point even earlier in time, where something that happened may have started the problem.

Earliest	Earlier	Now	Future

Look again at your list of variables. Visualize the behavior over time of each variable between the "Earlier" point and "Now." Then reconstruct each variable's "Earliest" behavior, if possible. Finally, imagine how the variable will behave in the future if nothing around it changes. These "thought experiments" can help you visualize the behavior of your variables over a broad span of time—to get the "big picture" of how they changed.[1]

Keep in mind that BOTs may be more free-form than other graphs you are used to seeing, especially if you work primarily with quantitative data. The lines are intended to indicate *qualitative* patterns over time rather than precise values. Of course, these patterns may eventually need to be verified by *quantitative* data later in the process.

Sketching the Graph

Here are some guidelines for drawing a BOT graph once you've selected a time horizon:

1. Graph your key variables together on the same graph. That way, you can see the variables' interrelationships—parallel variations, opposite variations, and delayed effects.

   Even though the variables are measured in different units, BOTs reveal how *variations* in behavior of variables might be dynamically related.

2. Label the lines clearly. If possible, use different colors to draw each variable.

3. On the horizontal axis, write the time horizon, either the number of years covered or the dates.

4. *Optional:* If a significant event occurred during this time frame—for example, a massive marketing campaign that directly preceded a jump in sales—draw and label a vertical line on the graph to show when it occurred.

   NOTE: If you find that a series of significant events associated with variables in the graph have occurred, you may have identified another variable to include in your graph. For example, suppose your graph showed that about every three years, a large percentage of people left the company voluntarily. In this case, you might add "Resignations" as a new variable and graph it alongside your other variables.

Using Your BOT Graph to Build Testable Hypotheses

Once you've sketched your initial BOT graph, the next step is to hypothesize about how the variables' behavior might be interrelated. This step often leads to ideas for new variables, additional BOT graphs, and yet more theories about how all the variables are connected. As with earlier stages of this process, at times you may feel as if you're "going backwards," but in fact it's this iterative quality of systems thinking that makes it so valuable for generating insights.

---

1. See Appendix F: Additional Resources for information about computer simulation. These programs can actually simulate the behavior over time of your chosen variables—much faster and more accurately than you can draw them out on paper!

In the A-to-Z semiconductor story, the managers selected a time horizon, set up their first graph, and drew lines representing the behavior of each variable (Figure 3.2, "A-to-Z's Performance Over Time"). Then, as they continued examining the problem, they asked questions, hypothesized about relationships among the variables, and drew additional graphs.

A-to-Z's initial graph depicted the behavior of sales, the sales force, profits, and new product releases. Observing that while sales were rising, profits were falling, the A-to-Z managers hypothesized that the problem might lie in the relationship between the total number of new products and the unit cost of carrying new products. Their second hypothesis theorized that the problem might stem from the relationship between the number of low-revenue new products and the level of the average selling price. These new hypotheses led them to create their second graph, shown in Figure 3.3, "Pressure on New Product Development."

Notice that the variables in this second graph are subsets or refinements of the original variables. These new variables were identified as the managers collected data, discussed the problem among themselves, and created their first graph.

As you try your hand at drawing BOTs in the learning activities that follow, be aware that it can take more than one attempt to identify the problem and the variables and to create a graph. The effort will pay off, however. Once you've completed these steps, you'll be ready to draw a picture of the system structure that has been generating the patterns you see on your BOTs: the causal loop diagram, discussed in Section 4.

FIGURE 3.2

A-to-Z's Performance Over Time

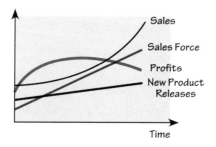

Sales revenue at A-to-Z had risen every quarter for the past 10 years, but profit growth had been falling for the last several quarters. Meanwhile, new product launches and the size of the sales staff were increasing every year.

FIGURE 3.3

Pressure on New Product Development

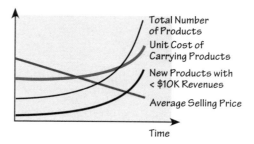

When A-to-Z's managers looked at their products' sales over time, they realized that although the number of products had been growing steadily, the cost of carrying the products was growing at an even faster rate. Drawing this BOT helped identify a self-perpetuating cycle, in which increasing revenue pressures put pressure on new product development to pump out new products that were easier to develop and launch, rather than innovative and potentially more profitable ones.

LEARNING ACTIVITIES

After trying the learning activities below, compare your responses with those in Appendix B. Don't worry if your responses look different from those in the appendix; there's no one right "answer" in a systems thinking analysis. These activities are meant mainly to get you started thinking about formulating problems, choosing variables, and graphing behavior over time.

Activity 1   THE PROBLEM WITH PRICE PROMOTIONS

Purpose: To identify the central problem in a complex situation

Outcome: A statement of the central problem in the case study
A description of a deeper problem that may lie beneath the central problem described in the case

Instructions: Read the case below and then answer the questions that follow.

Case Study ➤   ***The Problem with Price Promotions***

In the early nineties, a slowdown in U.S. population growth translated into smaller annual increases in consumer consumption, particularly of food products. Moreover, manufacturers' product innovation slacked off, and companies had trouble distinguishing their brands in meaningful ways other than through price. Their response: Offer price-cutting promotions to boost sales.

Manufacturers and retailers acknowledged that continual promotions could erode brand image and encourage consumers to shop solely on price. Also, companies could become dependent on short-term promotions to pump up sales numbers. Furthermore, manufacturers' promotions gave supermarkets especially great power, for they controlled the promotions on food products. Supermarkets could demand a wide range of subsidies, including fees for prime shelf space and money to pay for promotional material and newspaper ads.

The focus on price promotions spawned practices such as "forward buying," in which a supermarket bought more of a discounted product than it planned to sell during a promotion and then boosted its profit margin by selling the rest of the product at the regular price. Some supermarkets also "diverted" some of their low-priced shipments at a slight markup (but still well below the wholesale list price) to supermarkets outside the promotional area. For manufacturers, this meant that a large percentage of discounts intended for consumers wound up in retailers' pockets instead.

QUESTIONS

1. What are all the problems described in the case? List as many as you can.

2. From the point of view of the food manufacturers, what would you say is the overarching problem that includes many of the specific problems you named in Question 1?

3. Is there an even deeper problem behind the one you named in Question 2? If so, what is it?

ACTIVITY 2   THE CASE OF THE ENERGY DRAIN

Purpose: To practice drawing a behavior over time graph

Outcome: A statement of the overall problem in the story
A list of key variables
A graph of those variables' behavior over time

Instructions: Read the case below and then answer the questions that follow.

Case Study ➤   ***The Energy Drain***

It's 6:00 AM on Monday. The alarm blares, jolting you out of bed. You shuffle down to the kitchen to grab a cup of coffee. A few gulps and . . . ahhh. Your eyes start to open and the fog begins to clear.

10:30 AM, time for the weekly staff meeting. "I feel so groggy," you think. "I gotta have something to keep me awake through this one." You pour yourself another cup of coffee and head for the conference room.

Noon, and you're chatting with your colleagues at a quick lunch break. Someone refers to an article in the newspaper about fashion models' fitness routines. "Honest, it said that those high-priced runway models have to really watch it on caffeine. The way they keep their energy up is daily exercise and lots of sleep . . . 'beauty sleep,' I'll bet!"

Comments fly about who has time for daily exercise, getting paid to work out, and so on.

3:30 PM, you're feeling the mid-afternoon energy slump. You head to the crowded coffee cart to get another cup. "I really ought to cut down on this stuff," you comment to your friend in line. He nods. "I'm a five-cup-a-day guy, myself," he confesses. "I just can't give it up."

QUESTIONS

1. What's the problem in this story?

2. What are the three or four most important variables in the case?

3. What is the behavior of those variables over time? Graph them in the space below.

4. What do you observe about the behavior of the variables? For example, do any of them increase or decrease steadily over time? Are there dramatic changes (sudden upswings or plunges)? Do any of the variables seem to go through a cycle, as indicated by up-and-down patterns? Do any of them hold steady?

5. Do you observe any relationships among the behavior patterns of the variables? If so, what? For example, does one variable seem to rise or fall, followed by another variable's rising or falling?

ACTIVITY 3 THE CASE OF THE AUDIO-ELECTRONIC ROLLER COASTER

Purpose: To practice drawing a behavior over time graph

Outcome: A statement of the overall problem in the story
A list of key variables
A graph of those variables' behavior over time

Instructions: Read the case below and then answer the questions that follow.

Case Study ➤ ***The Audio-Electronic Roller Coaster***

AudioMax Corporation was on its way up. During the mid-1980s, the company enjoyed rapid growth and rising revenues. A manufacturer of special audio-electronics products, AudioMax served a growing core of clients who were willing to pay a higher price for better sounding, better quality innovations. AudioMax's products were well received because experts said they really did sound better. Initially AudioMax focused on maintaining a high degree of technical innovation, and this strategy generated a steady stream of new products.

Unfortunately, problems associated with AudioMax's subsequent rapid growth soon led the company into financial trouble. The company's CEO, Diane Schuster, had promised the investment banking community that immediately after going public in 1990 the company would beef up its management team. However, Schuster didn't bring in new people from the outside until 1991. When she finally added experienced management talent who could improve AudioMax's corporate strategizing, Schuster found it increasingly difficult to keep the company focused on R&D. Corporate expenses exploded, and productivity plummeted.

Meanwhile, the company began losing market strength, partly owing to its resistance to introducing models with simple cosmetic changes. Because the company was not introducing enough new products to keep customers interested, its dealer base began to deteriorate. In 1994, AudioMax had only 230 dealers, half as many as in 1990. Net income skidded sharply after 1991, followed by a heavy sales slump in early 1993 (Figure 3.4, "AudioMax's Income Slide").

AudioMax's problems persisted. The company cut costs by laying off one-third of the work force, but when sales rebounded, it could not rebuild staff fast enough to meet the new demand.

During the second half of 1994, AudioMax tried to get back on track by developing and releasing 18 new products. But as these new product lines were developed, the core products became even more dated.

During 1995, AudioMax saw net income rebound and sales increase over 1994. This time the CEO hoped the quality improvements would make a lasting difference. But the continuing cycle of ups and downs that followed AudioMax's rapid growth suggest an uncertain future for the company.

FIGURE 3.4

AudioMax's Income Slide

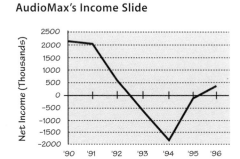

QUESTIONS

1. What was the long-standing, chronic problem facing AudioMax Corp.?

2. What two variables fed the initial steady growth in demand from AudioMax's customers?

3. Once AudioMax went public, what variable limited its ability to continue handling the growth in demand?

4. In the space below, graph the behavior over time of the two variables you identified in Question 2 and the one from Question 3. Then, add the behavior over time of two more variables: "Demand" and "Ability to meet customers' needs."

5. What do you observe about the behavior of the variables over time?

6. What relationships do you observe among the variables?

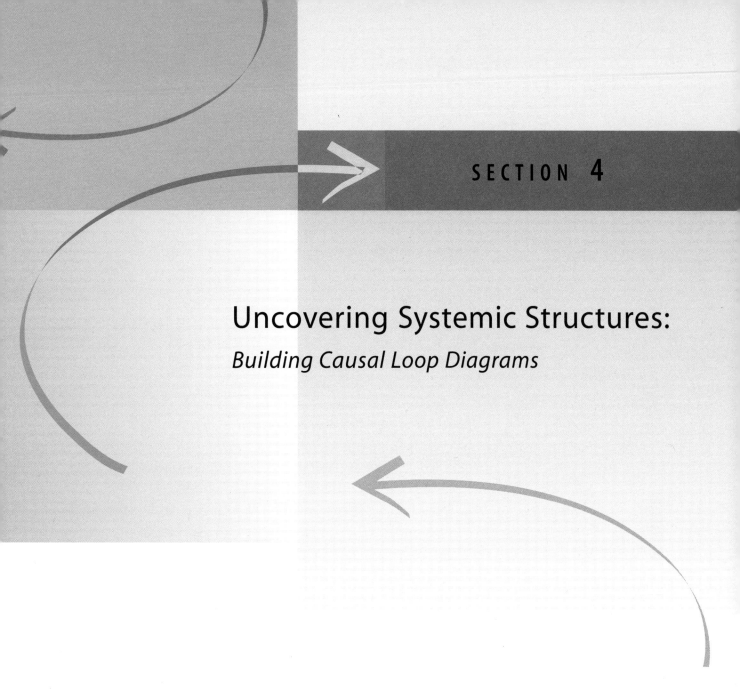

SECTION **4**

Uncovering Systemic Structures:

Building Causal Loop Diagrams

Now that you can graph patterns of behavior over time, you're ready for the next challenge: drawing a representation of the systemic structure itself. A graphic depiction of the structure—such as a causal loop diagram—lets you explore dynamic interrelationships among your variables that you may not have considered before. Sometimes, you can also see how parts of a system that are separated by location or time nonetheless might interact to generate problems. Finally, CLDs allow you to hypothesize about solutions to your problem and then test them, risk-free.

This section takes you through the next step in uncovering systemic structures: drawing CLDs.

ANATOMY OF A CAUSAL LOOP DIAGRAM

CLDs contain several components:

- One or more feedback loops that are either reinforcing or balancing processes
- Cause-and-effect relationships among the variables
- Delays

Read on for more details about these anatomical features!

Links and Loops

A CLD consists of two or more variables connected by links, which usually take the form of arrows. A closed circle of variables and links makes up a feedback loop, as shown in the two loops in Figure 4.1, "Links and Variables."

Relationships Between Variables: S's and O's

Not only does each link connect two variables, it has a sign that conveys information about the relationship between the variables.

In a CLD, you'll see at least two kinds of relationships between variables:

1. When variable A changes, variable B changes in the *same* direction.

Examples ➤   When the level of rainfall increases, the rate at which the grass grows also increases.

When hourly wages decrease, employee morale also goes down.

2. When A changes, B changes in the *opposite* direction.

Examples ➤   When level of fatigue increases, concentration wanes.

When the price of gold drops, the volume of gold purchasing goes up.

These relationships are shown on the loop diagram as either an "s" for a "same direction" change, or an "o" for an "opposite direction" change (Figure 4.2, "Links: Same or Opposite?").

We recommend that when you finish reading this book, you explore the more advanced tools of systems thinking beyond BOT graphs and CLDs.

FIGURE 4.1

Links and Variables

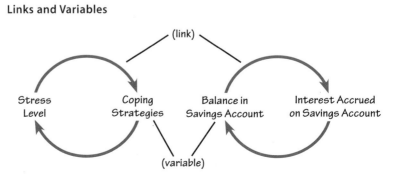

FIGURE 4.2

Links: Same or Opposite?

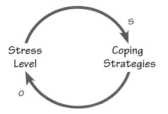

When stress level rises, the use of coping strategies also rises, ultimately reducing the stress level.

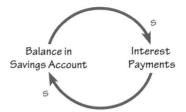

As your savings account balance increases, the interest you earn on the account also increases, further adding to your account balance.

Familiarizing yourself with the full palette of systems thinking tools, listed in Appendix C, is the best way to grasp the nuances and complexities of system behavior.

Reinforcing and Balancing Feedback Loops

Where do s's and o's lead us in interpreting feedback loops? We can think of feedback loops as closed circuits of interconnection between variables, and as sequences of mutual cause and effect. The links between each variable show how the variables are interconnected, and the signs (s or o) show how

A CLOSER LOOK AT S AND O NOTATION

The s and o notation method has some inherent peculiarities that require us to take a closer look at how we use causal loop diagrams. In this book, we use s's and o's, but in many other publications on systems thinking tools, you'll see "+" in place of "s," and "−" in place of "o." A lot of people find the s and o method easier to grasp than + and −, because the idea of "same" and "opposite" seems less confusing than "positive" and "negative." There is still much debate within the field of systems thinking about which notation method is more accurate.

Regardless of the notation method used, causal loop diagrams are valuable because they help simplify the dynamics of the system they depict, and they make complex dynamics accessible for newcomers to the field. They provide an important starting place for learning about systems behavior, and can inspire you to go on and master other systems thinking tools.

Any time you simplify something, however, you risk introducing inaccuracies, and CLDs are no exception. To see what we mean, look again at the savings-account loop in Figure 4.2. The s's make the most sense if you imagine each variable *increasing:* As interest goes up, the account balance goes up, gener-

ating even more interest, and so on. But what if the variables *decrease*? If the account balance goes down, interest payments go down, too. But as interest payments go down, the account balance doesn't go down, as the "s" would indicate (change in the same direction). Instead, the balance experiences what we call a "proportionate" *decrease* in the rate at which it *increases*. In other words, the account balance grows more slowly than before if interest payments go down.

This is tricky, but to be as accurate as possible, we need to acknowledge a distinction between a *proportional* change in the same direction, and a *direct* change in the same direction. Because causal loop diagrams present a simplified picture of system behavior, the notation used cannot capture this distinction. This doesn't mean we should abandon CLDs—they are a valuable tool for gaining insights into a problem, especially when a group works together on them. However, we do need to pay careful attention to when the signage of s's and o's can lead to confusing dynamics, and address these cases explicitly.

FIGURE 4.3

Employee-Supervisor Reinforcing Loop

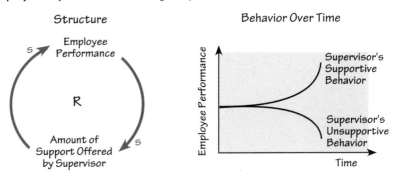

Reinforcing loops compound change in one direction with even more change. For example, encouragement from a supervisor can enhance an employee's performance, while critical or unsupportive behavior can lead to poor employee performance over time.

the variables affect one another. In this sense, CLDs are like simplified maps of the connections in a closed-loop system of cause-and-effect relationships.

Every feedback loop depicts either a reinforcing process or a balancing process. In fact, these two kinds of loops are the building blocks of any dynamic system structure, and they combine in an infinite variety of ways to produce the complex systems at work within and around us.

Reinforcing Loops: The Engines of Growth and Collapse

Reinforcing loops can be seen as the engines of growth and collapse. That is, they compound change in one direction with even more change in that direction. Many reinforcing loops have a quality of accelerating movement in a particular direction, a sense that the more one variable changes, the more another changes. For this reason, these loops are known as virtuous or vicious cycles—depending on the impact of the change! In causal loop diagrams, reinforcing loops are designated with an "R."

Figure 4.3, "Employee-Supervisor Reinforcing Loop," shows how a reinforcing loop can create either a virtuous or a vicious cycle.

Here's how the structure in Figure 4.3 works: Imagine that you're a supervisor who works with an employee named Rex: The more you support Rex (giving advice when he asks, showing appreciation for his contributions, and so forth), the more his performance improves. In some cases, the dynamic could start with improvements in Rex's performance that motivate you to act more supportively.

On the other hand, the less supportive you seem to Rex, the more his performance may slump over time. And as his performance erodes, your supportive behavior drops off even more.

In both of these scenarios, a reinforcing dynamic drives change in one direction with even more change in the same direction.

Sometimes you can detect a reinforcing loop at work simply by sensing exponential growth or collapse. You can also spot a reinforcing process in behavior over time graphs. When you see a steady upward or downward line, or an exponential upward or downward curve, a reinforcing structure is likely involved.

FIGURE 4.4

Employee-Supervisor Reinforcing Loop

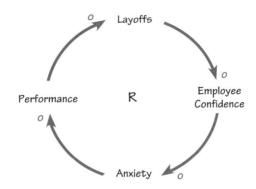

Increasing layoffs leads to a decrease in employee confidence. Dwindling confidence then leads to an increase in anxiety, and a decrease in performance—which leads to even more layoffs. This reinforcing loop has four o links.

A visual way to spot an R loop is to count the number of o's in a CLD. If there is an even number of o's (or no o's), then the loop is reinforcing. Although this is a handy method, you should always read through the loop and tell its story, to double-check that the label is accurate.

Note that reinforcing loops do not have to be made up of all, or even any, s links. In the loop in Figure 4.4, "The Reinforcing Power of Layoffs," for example, a company's increasing layoffs lead to decreasing employee confidence. The dwindling of confidence in turn leads to a rise in employee anxiety, which leads to a drop in performance, and then a further increase in layoffs. This diagram has four o links.

Balancing Loops: The Great Stabilizers

Nothing in life grows forever. There are other forces—balancing loops—that tighten the reins on those wild reinforcing loops. Balancing loops try to bring things to a desired state and keep them there, much as a thermostat regulates the temperature in your house.

Whereas the snowballing effect of reinforcing loops destabilizes systems, balancing processes are generally stabilizing or goal seeking. They resist change in one direction by producing change in the opposite direction. In CLDs, balancing loops are designated with a "B."

There is always an inherent goal in a balancing process, whether the goal is visible or not. In fact, what "drives" a balancing loop is a gap between the goal (the desired level) and the actual level. As the discrepancy between the two levels increases, the dynamic makes corrective actions to adjust the actual level until the gap decreases. In this sense, balancing processes always try to bring conditions into equilibrium.

An example of a balancing loop at work in a manufacturing setting would be the maintaining of parts inventory levels between production stages, as shown in Figure 4.5, "Inventory Control Balancing Loop." The system maintains a desired inventory level by adjusting the actual parts inventory whenever there are too many or too few parts in the warehouse.

FIGURE 4.5

Inventory Control Balancing Loop

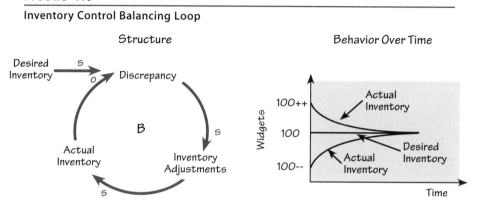

Balancing loops try to bring a system to a desired state and keep it there. In an inventory control system, the desired inventory is maintained by adjusting the actual inventory whenever there is too much or too little.

The inventory control structure works as follows: The company wants to maintain a certain level of parts inventory. When the actual inventory increases and exceeds the desired level, the gap between desired and actual inventory levels increases. This gap sends a signal to the company to adjust inventory levels by holding back on new parts orders. The adjustments bring the level of actual inventory back in line with the desired level, and the discrepancy between actual and desired shrinks or disappears.

The level of actual parts inventory can also drop *below* the desired level. Once again, the discrepancy between desired and actual increases, and the company orders more parts. Those adjustments bring the level of actual inventory in line with the desired level, and the discrepancy between the two is once again reduced.

How can you detect a balancing loop at work? One way is to watch for the goal-seeking behavior that characterizes a balancing process. In your organization, for example, if certain conditions keep coming back to some kind of "norm," no matter what anyone does, then a balancing process is likely at work. Similarly, if conditions seem to resist change, if growth falters or never quite starts, or if unproductive behavior never gets dropped, then a strong balancing dynamic is likely present.

In a BOT graph, you can spot a balancing process by the tell-tale rising and falling lines that over time always return to the center or "norm." You can also see it in growth or decline trends that eventually flatten.

A visual way to detect a balancing loop is simply to count the number of o links in the CLD. An odd number of o's indicates a balancing structure. As with reinforcing loops, however, you should still trace the story of the CLD in addition to counting o's, to ensure that the diagram is accurate.

Reinforcing and Balancing Loops Combining: The Building Blocks of Complex Behavior

To see how reinforcing and balancing loops can combine to form more complex behavior, let's revisit the employee-supervisor feedback loop in Figure 4.3. Clearly, Rex's performance will not improve indefinitely just because you're supportive. For example, poor Rex may be putting in longer

FIGURE 4.6

Employee-Supervisor Reinforcing and Balancing Loops

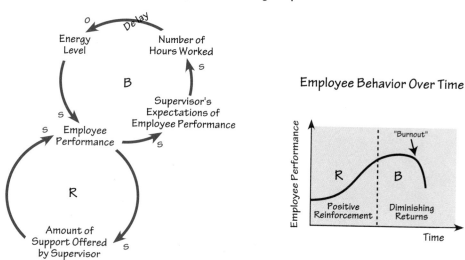

and longer hours in order to continue impressing you. Over time, Rex's long work hours may start wearing down his energy level (Figure 4.6, "Employee-Supervisor Reinforcing and Balancing Loops").

If this trend continues, the impact of your supportiveness will eventually be eclipsed by the drain of working long hours. Rex's performance improvements will gradually be offset by the effects of "burnout," until finally the balancing loop connecting energy level and hours worked dominates the structure. At this point, Rex's performance will either hit a plateau or actually decline.

The key point to remember is that all dynamic behavior is produced by a combination of reinforcing and balancing loops. Behind any growth or collapse is at least one reinforcing loop, and for every sign of goal-seeking behavior, there is a balancing loop. A period of rapid growth or collapse followed by a slowdown typically signals a shift in dominance from a reinforcing loop that is driving the structure, to a balancing loop.

Delays: The Hidden Troublemakers

In many systemic structures, delays play a hidden but important role. Delays themselves are neither good nor bad; it's how we humans handle them that determines whether they'll cause trouble. There are several ways in which we can fail to take delays into account: We can take too long to perceive feedback, to measure results, to decide how to respond to results, and to implement solutions. Two of the most insidious effects of delay stem from misperception. People often fail to take delay into account at all, or to realize that it even exists. We often have contrasting ideas, too, about how things in an organization work and how long developments take to unfold.

In a CLD, delay is depicted as a pair of lines (//) or the word *Delay* crossing the appropriate link (Figure 4.7, "Delay").

Delays are important to notice because they can make a system's behavior unpredictable and confound our efforts to control that behavior. For example, let's say consumer demand for sugar-free breakfast cereals is rising,

FIGURE 4.7

Delay

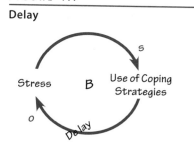

It can take time for coping strategies to begin reducing stress.

resulting in higher prices for these kinds of cereals. This heightened demand presents the Good Eats company with an opportunity to earn higher profits by producing more of the cereal and selling it at the new prices. The company decides to take advantage of the situation by expanding its capacity to produce the cereal.

But it takes time to expand. The length of the delay depends on how close Good Eats already is to capacity and how quickly it can grow. While the company is expanding, the gap between the overall supply of sugar-free cereals and the demand for them widens, and prices shoot even higher. The spiralling increase in prices drives Good Eats and its competitors to beef up their production capacity even more.

As the supply of the cereals eventually expands and catches up with demand, prices begin to fall. Supply overshoots demand, driving prices down further. By this time, Good Eats and other cereal manufacturers have overexpanded their production capacity. When prices drop low enough, sugar-free cereals will become even more attractive to consumers, and demand will pick up once more, starting the cycle all over again.

By understanding the relationships between the length of time it takes to increase capacity and the delay between changes in price and level of consumer demand, Good Eats and other cereal manufacturers might gain a better understanding of the implications of these delays. With this awareness, they could better prepare themselves for the inevitable oversupply that results from this all-too-common dynamic.

BUILDING A CAUSAL LOOP DIAGRAM

To draw a causal loop diagram, you want to pick up where you left off in Section 3 (drawing behavior over time graphs). To refresh your memory, here's how we recommend creating BOTs:

1. Formulate the core problem.

2. Tell the story of the problem behavior.

3. Choose the key variables you want to work with.

4. Name the variables precisely. Don't forget to:

   Use nouns or noun phrases.
   Be sure your variable name fits into phrases such as "level of" or "size of."
   Use a neutral or positive term whenever possible.
   Include intangible variables, such as morale, where appropriate, as well as tangible variables.

5. Graph the variables' behavior over time.

6. Hypothesize about how the variables might be interrelated.

There are several methods for moving beyond these steps to draw a CLD. Let's start off with a new case and then follow the steps up to drawing a BOT graph. Then, we'll describe three ways to use the graph to create a causal loop diagram.

Case Study ➤   ### *The Case of the Collapsing Banks*

Throughout its history, the United States has suffered periodic rashes of bank failures. During these episodes, depositors seemed to lose confidence in a bank and began withdrawing their funds. If word of this worry got around, more and more depositors lost confidence, and more and more funds were withdrawn from banks. Eventually, the volume of these withdrawals threatened the solvency of the bank, and when bank funds fell too low, the bank failed.

Worse yet, the failure of one bank could trigger a rash of other bank failures. Over the course of several months, depositors at other banks got nervous when they heard about the failure of the first bank, whether they had any reason to worry about their own banks or not. So they withdrew their funds from *their* banks, and, if funds got low enough, these banks, too, lost solvency and failed.

Here's how you might produce a BOT graph of this story's variables:

1. Formulate the Problem

To formulate the problem, ask yourself, What was going on? As we see it, the problem is that many banks were failing over the course of several months.

2. Tell the Story

The story, in brief, is that as depositors lost confidence in their banks, they withdrew their funds, and the banks began failing in a kind of domino effect. As more and more banks failed, depositors lost even more confidence and withdrew yet more funds. Then, even more banks failed.

3. and 4. Choose Your Key Variables, and Name Them Precisely

The significant variables that we detect in the story are:

> Bank failures
> Bank solvency
> Funds withdrawals
> Depositors' confidence

FIGURE 4.8

The Bank Story BOT Graph

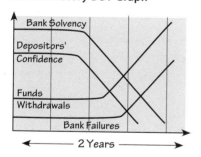

5. Graph the Key Variables' Behavior Over Time

Figure 4.8, "The Bank Story BOT Graph," shows how we see the key variables behaving over time.

Drawing the Causal Loop Diagram: Method 1—Begin at the Beginning

In this method, you draw the CLD by beginning at the start of the story. *The Case of the Collapsing Banks* begins with depositors' loss of confidence. If you start with "Depositors' confidence" as variable (A), what comes next? Which variable is directly affected by the loss of confidence?

When depositors lost confidence, they withdrew their funds from the bank. Another way of putting it is to say that when depositors' confidence

FIGURE 4.9

The First Link

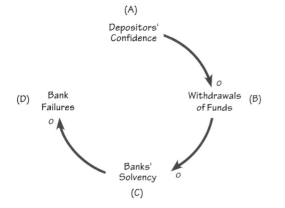

FIGURE 4.10

The Second Link

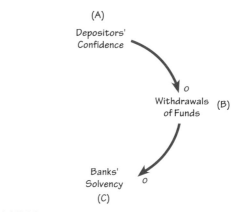

FIGURE 4.11

The Third Link

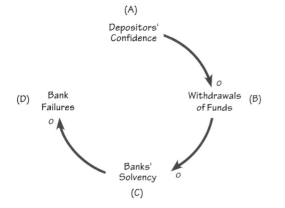

FIGURE 4.12

The Final Link

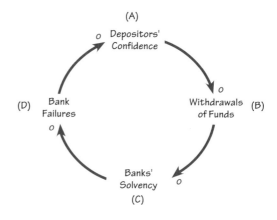

(A) dropped, withdrawals of funds (B) increased. That's the first link in your CLD, as shown in Figure 4.9, "The First Link."

When withdrawals of funds (B) increased, banks' solvency (C) decreased. Now there are two links (Figure 4.10, "The Second Link").

Finally, when banks' solvency (C) declined, bank failures (D) increased (Figure 4.11, "The Third Link").

What's the final piece of the structure that illustrates what drove the rashes of bank failures? Check the link between the increase in bank failures (D) and a further decline in depositors' confidence (A) (Figure 4.12, "The Final Link"). This crucial connection is what set the vicious cycle spinning!

Drawing the Causal Loop Diagram: Method 2—Work Backward

With this method, you start with the problem *symptom* and work backward to assemble the loop diagram. In this story, the problem symptom is bank failures (1). Of the identified variables, which one leads most *directly* to increasing bank failures (1)? Our answer is: decreasing solvency (2), as shown in Figure 4.13, "Link 2."

FIGURE 4.13
Link 2

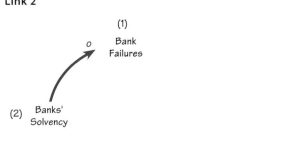

FIGURE 4.14
Link 3

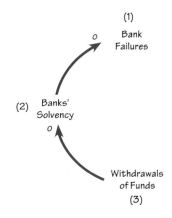

FIGURE 4.15
Link 4

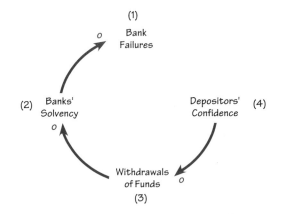

FIGURE 4.16
The Final Link

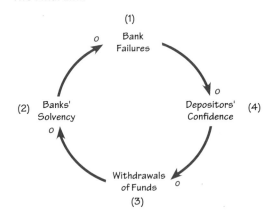

Withdrawals of funds (3) then lead directly to decreasing solvency (2) (Figure 4.14, "Link 3").

Increasing funds withdrawals (3) resulted from declining depositor confidence (4) (Figure 4.15, "Link 4").

Finally, what is the connection, if any, between declining depositor confidence (4) and the rising number of bank failures (1)? Self-evident, you say—the more bank failures, the lower depositors' confidence level (Figure 4.16, "The Final Link").

Drawing the Causal Loop Diagram: Method 3—Go Back and Forth

Most real-life systems stories are not so neatly composed and simple. Therefore, you may find that using a "back and forth" method is most helpful for building causal loop diagrams. In this method, you move back and forth through the sequence of the story, using bits and pieces of methods 1 and 2 to piece together your loop diagram. This method is particularly useful in diagramming complex stories that have many variables and loops.

You may also want to immerse yourself in several versions of the story. In a group effort, especially, just the act of identifying the variables, graphing their behavior over time, and describing their causal relationships can bring out the diversity of knowledge, perspectives, and opinions within the group. In some cases, building the causal loop diagram becomes secondary to exploring the breadth of views among team members.

So, although the "back and forth" method is time consuming and can seem messy, it is especially valuable when you don't have full knowledge of the system. In the end, it may even turn out to be the most fruitful method of all. Whichever method you use, the key is to stick with tracing a loop completely through before diverging onto other branches of the diagram. Otherwise, you may quickly find yourself in a mess of partially connected links with no clear focus. It takes discipline to stay with closing the current loop rather than pursuing what may look like more interesting paths that can come up in the diagramming process. But be patient—you can get to the other ones *after* you close the current loop.

Guidelines for Building Loop Diagrams

As you try the learning activities at the end of this section, keep the following guidelines in mind for building CLDs:

1. Remember to:

    • Use links and arrows to show the direction of the variables' cause-and-effect relationships.

    • Mark the links with s's and o's to show the nature (same or opposite) of the link.

    • Label the center of every loop with either an R for "reinforcing" or a B for "balancing."

2. If a variable has multiple consequences, try lumping them into one term while finishing the rest of the loop. For example, increasing stress may lead some people to increase how much they exercise, others to increase their alcohol consumption, and others to stay at work longer. All three of these consequences can be grouped together as "Use of coping strategies" (Figure 4.17, "Grouping Variable Names"). You can "unpack" these later when you're ready to explore the importance of specific strategies.

FIGURE 4.17

Grouping Variable Names

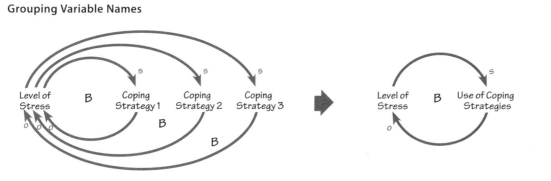

FIGURE 4.18
_____

Short Term and Long Term

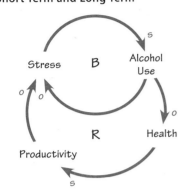

FIGURE 4.19
_____

Inserting a Variable

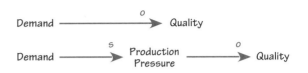

3. Almost every action has differing long-term and short-term consequences. To show increasingly longer-term consequences or side-effects, add new loops or links to your diagram. For example, in Figure 4.18, "Short Term and Long Term," increasing stress may lead to more use of alcohol, which, in the short run, can diminish the sense of stress. However, over time, the increased use of alcohol begins to destroy health, which leads to reduced productivity and worsens stress.

4. If a link between two variables is not clear to other people and requires a lot of explaining, try redefining the variables or inserting an intermediate variable to clarify the connection.

   For example, it may be clear to you that increasing demand for a consumer product could lead to reduced quality of the product. However, this simple expression of the connection may not be obvious to others. If you show that increased demand *first raises production pressure,* the resulting decline in quality becomes more understandable (Figure 4.19, "Inserting a Variable").

5. Check the reasoning behind your causal loop diagram by going around the loop and telling the story depicted by the links and the s's and o's. For example, the loop in Figure 4.20, "Telling the Story," depicts the following: "When the level of quality goes down, the gap between desired quality and actual quality gets larger. When that gap gets larger, actions to improve quality intensify. Eventually, after actions to improve quality have intensified, the level of quality goes back up."

FIGURE 4.20
_____

Telling the Story

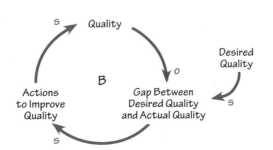

ABOUT MULTILOOP DIAGRAMS

Many CLDs contain multiple loops. In real life, most systems are so complicated that a diagram of even a portion of them can look like a giant plate of spaghetti! We don't encourage you to tackle this level of complexity just yet, but you may want to know a little bit now about how to interpret multiloop CLDs. The following tips are meant to get you started interpreting complex loop diagrams. As you continue your adventure into systems thinking after reading this book, you'll have lots more opportunities to play with multiloop CLDs.

To "read" the story in a multiloop diagram, begin by tracing through one loop, usually the one containing the variables from the first part of the story that the CLD depicts. For example, in Figure 4.21, "A Multiloop CLD," the central loop contains the variables "Use of outside supplier parts," "Costs," "Margins," "Net income," and "Financial pressure."

To read subsequent loops in the diagram, choose the variable in the central loop that first branches away from the central loop (in this case, "Use of outside supplier parts"). Continue the story into the next loop ("Premium brand image"), then come back around to the joining point ("Net income"). Finally, continue tracing through until you get back to "Use of outside supplier parts."

FIGURE 4.21

A Multiloop CLD

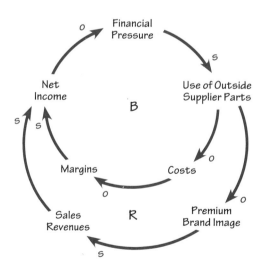

After completing these learning activities, compare your responses with those in Appendix B.

ACTIVITY 1   THE CASE OF THE PLATEAUING PROFITS

Purpose: To learn to distinguish reinforcing and balancing processes

Outcome: Awareness of the distinguishing characteristics of reinforcing and balancing loops

Instructions: Read the case study below and then answer the questions that follow.

Case Study ➤ **Plateauing Profits**

Medicorp, an HMO, opened its doors for business in 1985. During its first five years, it advertised heavily, and its customer base climbed by an impressive 35 percent annually. Medicorp was the first HMO in the area to offer its customers an unusually broad selection of physicians for the same price as other HMOs' policies, and as word of mouth intensified, more and more customers flocked to Medicorp. At Medicorp's five-year-anniversary mark, the management team celebrated in high spirits, confident that Medicorp's spectacular success was bound to continue.

In the early 1990s, however, a few other HMOs, attracted by Medicorp's success, moved into the area and began offering similarly appealing coverage policies. As the mid-1990s rolled by, Medicorp noticed a slowdown in the rate of growth of its customer base. The 35-percent annual rise began dwindling to 25 percent and then 15 percent, and then it leveled off at a lukewarm—though respectable—10 percent.

QUESTIONS

1. Considering just the story, and without listing variables or drawing BOT graphs, which kind of process—reinforcing or balancing—do you suspect was at play during Medicorp's *early* years? Explain your answer.

2. Do you sense a shift in process during Medicorp's history? If so, which kind of dynamic—reinforcing or balancing—prevailed near the *end* of Medicorp's story? Explain your answer.

3. Draw a simple BOT graph that shows just the pattern of the growth of Medicorp's customer base. Based on your graph, which kind of process—reinforcing or balancing—do you think describes the overall Medicorp story?

4. What do you think made Medicorp's growth hit a plateau?

Activity 2   **THE CASE OF THE COLLAPSING BANKS**

Purpose: To learn to distinguish reinforcing and balancing loops

Outcome: Awareness of the defining characteristics of reinforcing and balancing loops

Steps

1. Look again at The Case of the Collapsing Banks on page 59. By simply reviewing the story (don't count the s's and o's in the finished CLD), what kind of dynamic—reinforcing or balancing—do you sense at work?

2. Look again at the finished CLD for the collapsing-banks story (Figure 4.12 on page 60), and count the o's in the diagram.

3. What kind of loop is the collapsing-banks CLD? Explain your answer here and label Figure 4.12 with the appropriate letter: R or B.

Activity 3   THE "ORGANIC TO GO" STORY

Purpose: To contrast nonsystemic and systemic perceptions of a problem
To recognize how we think, talk, and act differently when we have a systemic view of a problem

Outcome: Insights into the possibilities offered by systemic thinking
Familiarity with tracing a story through causal loop diagrams

STEPS

1. Read "Organic To Go: Part One," below.

2. Answer the Part One questions, without turning to "Organic To Go: Part Two."

3. Read "Organic To Go: Part Two."

4. Answer the Part Two questions.

Case Study ➤   *Organic To Go: Part One*

About two-and-a-half years ago, an organic food retailer opened a new line of business—Organic To Go—that offered takeout health-food meals. Customers call or fax orders from a menu of snacks, salads, soups, sandwiches, and entrees, and then pick up their orders at a designated time. Customers may also walk in and order at the counter. Although there is a small eat-in area, most customers order takeout.

Organic To Go has grown vigorously, attracting lots of young employees. The entire staff has been amazed at the volume of customers that responded to the new concept. Everything about Organic To Go has grown—volume, customers, locations, staff, and menu. The owners are young, with a big vision, and they gladly make room for employees to rise into management ranks. Because they have opened new locations all over the city, employees have had many opportunities to learn about the business and to develop new skills. Just in the last year, eight new sites were opened, including two in outlying suburban areas.

In the past few months, Organic To Go's top managers have noticed some hesitation in the business's performance. The number of customers seems to be leveling off. Revenues are no longer booming. When site managers are consulted, they report some increase in a variety of problems—botched filling of orders, sloppy restocking, equipment and facility maintenance problems, and employee absenteeism. There's no clear trend, and different locations are struggling with different kinds of problems.

Figure 4.22, "Organic To Go Over Time," shows the behavior of the story's variables over time.

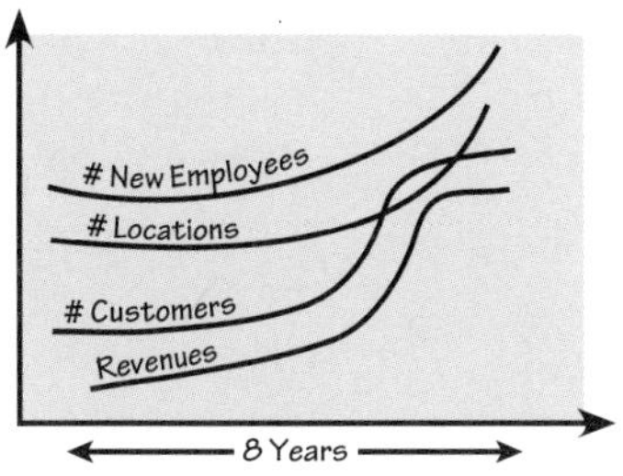

FIGURE 4.22

Organic To Go Over Time

QUESTIONS

1. How would you explain the troubling trends at Organic To Go?

2. Is there anything else the managers should take into account?

3. In your opinion, what should the Organic To Go managers do?

4. What typically happens in your organization when sales volume, number of customers, or profit margin slows down?

Note: In the story below, variables are followed by letter labels that match the variables in the accompanying causal loop diagram (Figure 4.23, "The Organic To Go Loop").

Case Study ➤   ***Organic To Go: Part Two***

Once Organic To Go was established and successful in its first location, its managers allowed themselves to dream about the future. They knew they were onto a "hot" idea. They could visualize how, as they opened new locations (A), copying the successful model created in the original "mother store" would help them increase their customer base (B). As growing sales volume (C) brought in more revenue (D), funds would become available for investment (E) in more locations. Organic To Go could be another Starbucks—bicoastal, maybe even international!

If we expand the loop as shown in Figure 4.24, "The Bigger Picture," however, we can see how problems began to arise at Organic To Go.

As the number of locations (A) increased, the number of new employees (1) increased. Fewer of those employees (2) had previously worked in the original "mother store." That meant that an increasing number of new employees didn't know the policies, operating procedures, and behavior expectations that had made the original business such a big hit with customers. As they tried to learn on the job, the new employees put an

FIGURE 4.23

The Organic To Go Loop

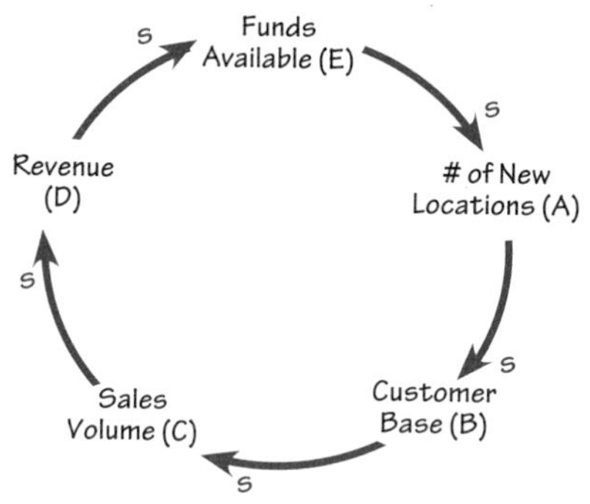

FIGURE 4.24

The Bigger Picture

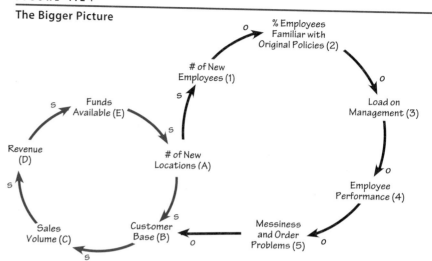

increasing load (3) on their managers. Sometimes the newer employees got frustrated and upset, and came late, left early, or took a day off without notifying their managers (4).

In addition, occasional problems with late or incomplete orders and with messy serving and eating areas (5) turned off some customers, who either came less often or stopped coming in at all. Some of those customers complained to their friends, so Organic To Go stopped attracting as much new business (B) as it had before.

QUESTIONS

1. Look again at the first Organic To Go causal loop diagram, in Figure 4.23 above. Does it represent a reinforcing or a balancing process? Label the loop with an R or a B to show your decision.

2. Look at the second causal loop diagram, in Figure 4.24 above. Does the new section of the diagram represent a reinforcing or balancing process? Label the section with an R or a B.

3. When you look at both the original vision and the expanded depiction of the story's actual outcome, do you see the issues at Organic To Go differently than you did at first? If so, how?

4. What suggestions would you offer the Organic To Go managers?

5. What factors in particular do you now think the managers should pay most attention to?

6. Think about using causal loop diagrams in your own organization when sales volume, number of customers, or profit margin slow down. In your opinion, how might you and your colleagues see these problems differently by using CLDs?

ACTIVITY 4   **THE CASE OF THE RESTRICTED REVENUES**

Purpose: To follow the steps for discovering the dynamic structure behind the story

Outcome: A causal loop diagram of the story

Instructions: Read the case study below and then answer the questions that follow. (The underlined sections in the case study are intended to help you see the story's key variables.)

Case Study ➤   **Restricted Revenues**

The year 1991 marked a series of firsts for PC-Plus, an established IBM-PC clone manufacturer. That year, PC-Plus experienced its first quarterly loss ($20 million), its first round of layoffs (1700 people), and the first departure of a chief executive. These were not the kinds of records PC-Plus was used to posting. In the late 1980s, the company created a sensation in the business world by growing to $1 billion in sales faster than any other American firm in history. But then the early 1990s saw PC-Plus losing market share to other clone makers and struggling to stay on top.

PC-Plus's initial strategy had been simple: Build IBM-compatible computers that cost about the same as the competition but that either performed better or offered extra features. PC-Plus's engineering strength, combined with its marketing savvy, jump-started its early success. The company could command premium prices by offering technologically sophisticated products. But in 1991, customers began perceiving PC-Plus's products as overpriced, and questioned the company's leadership role in engineering breakthroughs. In 1986, PC-Plus had leaped ahead of its biggest rival by bringing out the first IBM-compatible machine using a new, faster microchip. But in 1991, PC-Plus sat on the sidelines while three other clone makers announced their own new machines with an improved microprocessor. These competi-

tors asked customers why they should pay
PC-Plus's high prices—and the customers listened.
To the consternation of the PC-Plus management
team, the company's <u>revenue growth</u> threatened to
flatten during the late 1980s and early 1990s, as
Figure 4.25, "The Growth of PC-Plus," reveals.

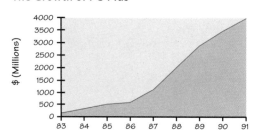

FIGURE 4.25

The Growth of PC-Plus

To maintain its success, PC-Plus needed to
approximate its competitors' prices without losing
its reputation for quality. It also needed to explain
in its advertising and marketing why buyers should
pay extra for a PC-Plus computer that boasted the same features as the competition. And it needed to get closer to the customers it used to keep at arm's
length. Management responded to each of those concerns with a variety of
measures: across-the-board price cuts up to 34 percent to compete more
directly on price; expanded distribution channels, including some computer
superstores and third-party resellers to increase accessibility; a reorganization
into two semiautonomous divisions to speed product development; and a
toll-free hotline to keep in touch with customers.

Though each move addressed PC-Plus's major shortcomings, they also carried risks. Even with the cuts, PC-Plus's prices remained high, <u>and the lower
margins ate into the company's profits</u>, which might have hurt its ability to
invest in further research and development. Some analysts worried that PC-Plus's plan to keep costs down by <u>buying more parts from outsiders would
hurt the company's premium-brand image</u>, forcing it to slash prices even
more to remain competitive. Expanded distribution channels might also have
sullied PC-Plus's image by making it appear to be "just another clone."

QUESTIONS

As suggested by the underlined sections in the story, the key variables we identified in
the PC-Plus story are:

Costs
Margins and Net Income
Financial pressure
Use of outside suppliers' parts
Premium-brand image
Sales revenues

Figure 4.26, "PC-Plus Over Time," shows these variables' behavior
over time. To figure out what structure drove PC-Plus's response to
the financial pressure of a slowdown in revenues, we suggest that
you work with the first four variables from the above list, which we
highlighted in Figure 4.26:

Costs
Margins and Net income
Financial pressure
Use of outside suppliers' parts

FIGURE 4.26

PC-Plus Over Time

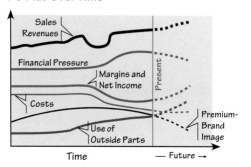

In the space below build a causal loop diagram of the structure, using these steps as guidelines:

1. The problem seems to be PC-Plus's increasing financial pressure (A).

   What variable (B) led to the increasing financial pressure? (Draw the link, and label it with an s or an o.)

2. Which other variable (C) is associated with (B) and caused (B) to change?

3. Which variable (D) did PC-Plus decide to change in order to affect (C)? (In other words, PC-Plus hoped that if they did more of (D), (C) would be reduced.)

4. To complete the loop, add a link between (A) and (D).

5. What process does this loop represent—reinforcing or balancing? Label your finished loop with an R or a B.

6. Check your loop for soundness. Be sure you drew all the links, with arrows to show the direction of movement, and that you have labeled all your links with s's and o's, to show how each variable changes in relation to the one that influences it. Go around the loop and tell the story.

Activity 5   **THE ALL-FOR-ONE COOPERATIVE**

Purpose: To practice the steps for discovering the structure behind a pattern of behavior

Outcome: A list of the story's key variables
A behavior over time graph of the key variables
A causal loop diagram that depicts the case story

Instructions: Read the case study below and then answer the questions that follow.

Case Study ➤   ### The All-for-One Cooperative

The All-for-One Cooperative is a stunning example of a community that has embraced and mastered business collaboration. The cooperative was founded several decades ago with funds raised from local townspeople to open a small factory that employed only 25 people. Today, All-for-One has more than 160 cooperative enterprises, employing more than 23,000 members. The member enterprises' actions are based on a single guiding principle: "How can we do this in a way that serves equally both those in the enterprise and those in the community, rather than serving one at the expense of the other?"

Interviews with members of the cooperative revealed the centrality of this shared vision between the business and the community. Trusting in their shared beliefs and values had encouraged the founders to collaborate closely to design the original cooperative. Reinforcement of their common beliefs had allowed newer, younger members to propose the changes required for All-for-One to adapt as it grew.

The experience of collaborative design supported various joint experiments in profit sharing, diversification of members' benefits, apprenticeships, and exploration of new markets. Although some of the experiments failed, the losses were spread across the entire association. As a result, the experimenters were encouraged to discuss their ventures at cooperative governance meetings and to publicize the causes of both success and failure to all cooperative members. Old timers and newcomers believe that this sharing of business knowledge has consistently reinforced the sense of shared vision and beliefs.

QUESTIONS

1. Briefly sum up the All-for-One story.

2. Identify four or five key variables from the story.

3. Graph the variables' behavior over time.

4. Decide how the variables interrelate, and draw a causal loop diagram to show the connections. Work in pencil, and don't be afraid to use that eraser! Label your loop diagram R or B.

5. Walk through the causal loop diagram, telling the story it depicts and checking it against the original story.

ACTIVITY 6   **THE PROBLEM WITH USED CDS**

Purpose: To practice all the steps of discovering the structure underlying a pattern of behavior

Outcome: A list of the story's key variables
A graph of the variables' behavior over time
A causal loop diagram of the story

Instructions: Read the case study below and then answer the questions that follow.

Case Study ➤   ***Used CDs: What Now?***

The compact disc, although quite expensive, has a huge advantage over the old LP: It lasts almost forever. This feature is highly attractive to consumers, but poses a concern for record companies because of the growth of the used-CD business. Indeed, the popularity of CDs has triggered an inevitable confrontation over the selling of used discs—a clash that pits the giant record companies against retail outlets. LP retailers complain that much of their used-CD inventory comes from record clubs or from free promotions. Although not illegal, this practice, the retailers argue, cuts into the sales of new CDs, diminishes their value, and deprives recording artists of royalties.

At the heart of the controversy over the sale of used CDs is the decision by a large record chain, SoundAlive, to carry half-priced used CDs alongside new ones. The issue is further complicated by the fact that SoundAlive has always been viewed as an industry innovator, opening the possibility that other large chains may start to carry used CDs to remain competitive.

Ironically, the actions of some record companies have actually contributed to the rise in used-CD sales. For example, when SoundAlive offered its customers a no-hassle return policy on CDs, some suppliers stopped accepting returns of opened CDs from retailers. Instead, they offered a 1-percent credit. With many returned CDs on their hands, SoundAlive found it more profitable to resell the discs at a discounted price than to return them to the record company.

QUESTIONS

1. What was the original problem that CDs posed for record companies?

2. What did the record companies do to overcome this problem?

3. What was the next problem or issue facing the record companies?

4. Identify a total of four key variables reflecting the original problem and the later problem.

5. Create a graph of the variables' behavior over time.

6. Draw a simple two-loop diagram that shows the original problem and the later problem, including how the later problem is linked to the original problem. (See "The Case of the Restricted Revenues," page 70, if you need help getting started.) Label your loops R or B.
   HINT: Draw the loop that shows the original problem first. Then add the loop that shows the subsequent problem.

Complex Systems

In the earlier sections of this book, you focused on simple causal loop diagrams, with one or two loops. These are fine for depicting relatively basic systems and even for capturing the most fundamental dynamics of a more complicated system. However, most systems we encounter at work, in our communities, or in news of current events are large, complex, and extremely difficult to diagram. For example, imagine trying to create a causal loop diagram of an entire corporation, a city, a national economy, or an ecosystem such as a forest. All of these are complex systems, and their diagrams would contain vast numbers of reinforcing and balancing loops and complicated interconnections among the loops.

This section outlines some of the inherent characteristics of complex systems[1] and then presents the story of ComputeFast—a company whose trials and challenges vividly demonstrate complex system behavior.

---

1. Adapted from Draper Kauffman, Jr.'s, *Systems 1: An Introduction to Systems Thinking* (Future Systems, Inc.), 1980.

CHARACTERISTICS OF COMPLEX SYSTEMS

Complex systems behave differently from simple systems and pose special challenges for systems thinkers. In action, a complex system appears to have many variables, many factors at play, and many semi-independent but interlocking components. In a diagram of a complex system, the dominance of the different feedback loops shifts, and the timing and length of delays vary. The diagram may also depict a number of structures that even seem to be in conflict with one another.

Here are some further defining characteristics of complex systems:

1.   Complex systems tend to be self-stabilizing.

A causal loop diagram of a complex system is likely to contain a great many balancing loops, each of which acts to keep some smaller component of the system in balance, or functioning close to some desired level. Picture all the legs of a centipede moving the insect toward its goal. Even if a few of the legs are broken or wobbly, the sheer number of them allows the creature to keep crawling forward quite smoothly.

This characteristic suggests why so many complex organizations resist change or improvement campaigns, and eventually return to status quo: All the balancing loops are designed to keep things the way the system originally intended them.

2.   Complex systems are or appear to be purposeful.

Complex systems contain numerous balancing loops, each of which attempts to maintain a desired level of performance or a goal. A complex system may contain a number of reinforcing loops, too, each of which serves to augment or diminish some kind of phenomenon within the system. In a complex system such as an organization, sometimes the goals of both the growth and the balancing processes are explicit and known to the people who make up the organization. But very often, they are contradictory, ambiguous, or implicit, and the system appears to function with a mind of its own.

3.   Complex systems, like simpler systems, are capable of using feedback to modify their behavior.

All systems use feedback to modify their behavior. This ability provides a key opportunity for change and growth within the system—especially if the feedback is explicit and accessible. For example, an organization has a better chance of improving its performance if it seeks to gather information about problems such as shipping delays or procedural complexities. For anyone studying systems, a good understanding of the structure and workings of a system makes it easier to take advantage of this important capability so as to catalyze change within the system.

4. Complex systems can modify their environments.

Because systems seek to fulfill a purpose and can modify their own behavior, it's not surprising that they can also modify their environments, the better to achieve their goals. These modifications may be subtle, such as pedestrians' wearing a diagonal path across a square of lawn, or quite bold, such as a development company's decision to build a suspension bridge or demolish a mountainside.

For anyone wishing to change their surrounding environment, a vital step is to identify the links between the system in question and its environment. With this awareness of how each system is part of a larger system, we can anticipate how changes we might make in one system will lead to changes in the system's environment.

5. Complex systems are capable of replicating, maintaining, repairing, and reorganizing themselves.

Franchises and branch offices are examples of how an organizational system replicates itself. However, because systems also change in response to their environment, even apparent clone organizations are likely to contain unique quirks or mutations.

In addition, organizations that are abruptly altered, let's say through a take-over or layoffs, often find ways to carry on their essential functions or reorganize themselves to continue pursuing their essential goals. All natural systems have this power to invent, reframe, learn, and adjust to their environments.

 ## THE STRENGTHS AND WEAKNESSES OF COMPLEX SYSTEMS

The more complex a system is, the greater its potential to process large quantities of information, learn quickly, and act flexibly. However, complex systems also have many subsystems to coordinate, and therefore more potential for things to go wrong. Complex systems often meet with four kinds of problems:

• Conflicting goals

• The centralization vs. decentralization dilemma

• Distorted feedback

• Loss of predictability

In a complex system, it is not uncommon for subsystems to have goals that compete directly with or diverge from the goals of the overall system. For example, an organization may advocate local decision-making, only to discover that some local decisions conflict with the goals of the overall organization. The organization may react by swinging toward centralized decision-making. Feedback gathered from small, local subsystems for use by larger subsystems may be either inaccurately conveyed or inaccurately interpreted. Yet it is this very flexibility and looseness that allow large, complex systems to endure, although it can be hard to predict what these organizations are likely to do next. Compare the endurance and vitality of a market economy, for instance, to a controlled economy.

COMPUTEFAST: THE STORY OF A COMPLEX SYSTEM

The story below—about a company that's thinking of expanding—highlights some typical behavior patterns of complex systems. We've also illustrated the ComputeFast case with several examples of the graphs and causal loop diagrams discussed in this book, to give you a sense of what more complicated CLDs look like. Perhaps most important, the story and diagrams show you how an actual organization—a complex system—might use systems thinking tools to begin to gain insights into its business issues and explore the many ramifications of its decisions.

Can ComputeFast Service Its Success?

In the world of mail-order computers, maintaining a balance between improving technology and developing new markets has always been a challenge. ComputeFast is facing this issue, experiencing increased competition at home while preparing to expand into new overseas and domestic markets. The company must find a strategy that encompasses this expansion—while still keeping its service quality high and its initial customer base happy.

ComputeFast took the lead in the U.S. mail-order PC computer business by combining low production costs with a customer base of small businesses and technically knowledgeable users. A "no-frills" corporate style also allowed the company to undercut competitors on price. These benefits have enabled ComputeFast to offer its customers an affordable, high-quality product along with good customer service.

Yet the company has to keep growing its revenues while maintaining quality and service—and not losing control of costs. For the first time since it opened its doors for business, ComputeFast suffered a drop in sales over the past year. The company's CEO attributes the decline to a backlog of orders from last December that inflated first-quarter results. Future drops in sales, however, may hint at a larger problem: declining customer service quality.

Finding and training local technical and assembly-line workers quickly enough to keep up with customer demand has been difficult. This staff shortage has worsened customer complaints of delayed deliveries and long waits on customer service phone lines. By adding 70 new phone lines and expanding its cadre of 180 technical support personnel, ComputeFast has managed to halve the time that customers must wait to speak to a technician. Still, some customers may spend up to six minutes on hold.

Improving Service vs. Expanding Markets: The Balancing Act

At the same time it is facing these problems at home, ComputeFast is planning to create a production facility in the Far East and market mail-order PCs in Europe. These expansion efforts mirror the company's early U.S. strategy of finding a production base that will keep initial costs low, thus allowing the company to provide quality service at a reduced cost. ComputeFast may find, however, that plans to expand overseas are harder to implement with a dissatisfied customer base at home.

Another part of the company's strategy for increasing its competitive position involves marketing to a broader domestic corporate base while expanding its market overseas, as shown in the reinforcing loop in Figure 5.1, "Expanding Markets."

Before it plunges into all these new plans, ComputeFast needs to ask itself some key questions:

- How fast should the company expand its technical support capacity compared to its revenue growth?

- Once a growing company identifies a capacity shortfall, it's easy to anticipate acquisition delays—but what are some of the *hidden* sources of delay that could catch ComputeFast by surprise?

- In thinking about the possible erosion of its performance standards as ComputeFast expands, the company also needs to acknowledge that standards may become obsolete. How might ComputeFast's current standards of service become inadequate?

- Is ComputeFast's goal of expanding into new markets, maintaining profit margins, and improving service quality an achievable one?

FIGURE 5.1

Expanding Markets

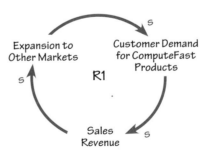

As customer demand for ComputeFast's products increases, revenues also increase, providing the necessary capital to expand into new markets and further spur demand.

The Problem of Success

In ComputeFast's situation, the current technical-support service problems stem directly from the company's sales success. Revenues skyrocketed from a mere $100,000 in 1990 (its first year) to $70 million in 1994, and topped $1.11 billion in 1996. Estimated revenues for 1998 are around $1.7 billion. As the number of units sold continues to climb, the number of customers requiring service also increases. This trend suggests that the company should expand its service capacity *at least* as rapidly as the growth in shipments. In terms of total number of employees, ComputeFast appears to have kept pace with revenue growth. From 1994 to 1996, the employee ranks grew almost fivefold, from 400 to almost 1,860.

But, as ComputeFast has learned, simply adding workers is one thing; finding people with the appropriate skills and experience is altogether different. In particular, acquiring and training service technicians has proved more challenging than the company may have first realized. Consequently, ComputeFast customers have been experiencing delivery delays and busy customer service phone lines. Over time, these annoyances will dampen demand, according to the balancing loop B2 in Figure 5.2, "Balancing Demand."

When a company experiences eroding service quality, halting the slide requires beefing up capacity as quickly as possible. Yet in making the decision to invest in capacity, the company must immediately recognize when performance actually falls off and when the needed capacity should be added. In many cases, the delay between recognizing declining levels of service and adding needed capacity widens because the company is unable or unwilling to face up to the inadequacy of its current efforts to halt the quality decline.

In addition, the challenge of keeping after-sales support on track is complicated by two unavoidable facts: When sales are growing linearly, the installed base is growing exponentially; and when sales are growing exponentially, the installed base is growing super-exponentially, as suggested in

FIGURE 5.2

Balancing Demand

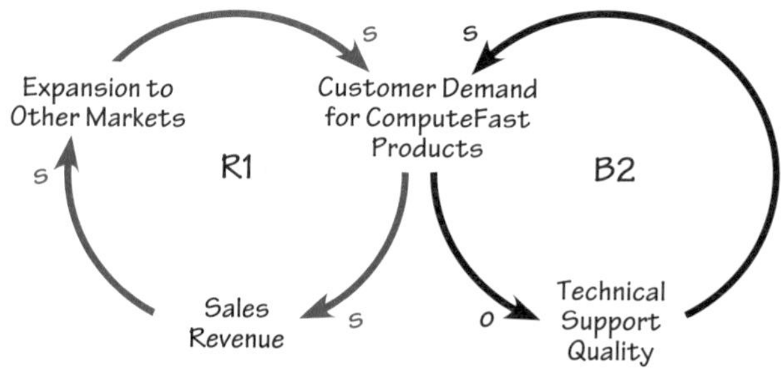

As customer demand increases, the quality of ComputeFast's technical support decreases, which reduces demand (B2). If the low demand persists, the customer base will be easier to service, and technical support quality will then increase.

FIGURE 5.3

Exponential Growth

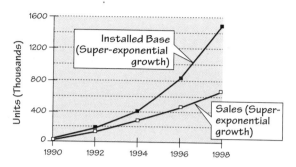

Figure 5.3, "Exponential Growth." This means that, to provide adequate after-sales support, the number of service technicians may need to grow at a rate much faster than the organization as a whole.

If ComputeFast is going on the assumption that the number of service technicians should grow at the same rate as sales, the company is likely to chronically underinvest and eventually will fail to provide customers with the level of support it offered in the past. This delay in recognizing true capacity needs may, in a self-limiting way, control ComputeFast's growth by stifling technical support quality (and keeping loop B2 in Figure 5.2 active). If the low service quality persists and depresses customer demand, the company will find it easier to service the lower demand. Higher quality service will reduce the quality gap, and the narrowing gap will likely lead to lower investments in capacity and, eventually, a lower technical support quality, as shown in loop B3 in Figure 5.4, "Underinvestment in Service." In this

FIGURE 5.4

Underinvestment in Service

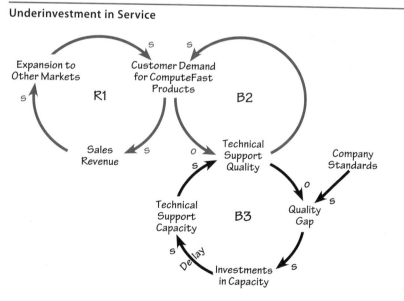

An increase in technical support quality will *reduce* the quality gap, which will ease the pressure to invest in capacity, resulting in even lower technical support capacity (B3). Loops B2 and B3 together form a figure-8 reinforcing loop of continual underinvestment in service capacity and declining customer demand.

diagram, loops B2 and B3 can spiral in a vicious figure-8 reinforcing loop that leads to lower demand and lower service capacity. When caught in this dynamic, the company may even continue to believe it's offering an "acceptable" level of service.

Even if ComputeFast recognizes the need to expand its technical support at an accelerated rate and is somehow able to hire and train all the people the company feels it needs, it can still fall victim to the figure-8 dynamic. This scenario is likely to happen if ComputeFast's performance standards, which drive its investment decisions, are not continually updated to fit the company's changing situation.

The value of a performance standard can still erode if the standard is never revised to meet current market conditions. In this case, the company will never achieve adequate investments needed to support the market's changing quality expectations. The result: Demand will decrease until it hits the number (and type) of customers who are satisfied with the company's traditional standard.

For example, the new corporate customers that ComputeFast is targeting will most likely have higher service needs and expectations than the company's original core market of technically savvy users. And people who are accustomed to the kind of service that other, larger companies offer may find ComputeFast's service level inadequate, even at its best. A push into such new markets may require a substantial redefinition of ComputeFast's notion of high service quality.

Investing to Expand

Improving customer service in the face of continuing rapid growth is a challenge in itself. Couple that with a strategy for entering foreign markets and targeting more established corporate buyers, while still preserving current profit margins, and the goals seem to be at cross purposes—at least in the short term as shown in Figure 5.5, "Pursuing Two Goals at Once."

Expansion into the overseas direct-mail market involves a complicated and costly web of marketing and shipping arrangements that will require major investments. At the same time, the new corporate buyers and the ever-expanding installed base at home will require further investment in technical support capacity. Both investments will increase costs and squeeze profit margins, for sales are not likely to rise very fast in the early stages of overseas expansion. If both investments are pursued aggressively, profits will suffer and ComputeFast will feel pressured to cut back (loops B4 and B5 in Figure 5.5). However, if the company pursues technical support investments less aggressively, it may still be able to target the new markets and maintain profit margins. The risk, of course, is that service quality may suffer.

Here's one possible scenario for ComputeFast to consider: proceed with its expansion plans while keeping a careful eye on the bottom line. The company can strengthen its technical support incrementally, but only after it receives complaints about inadequate service quality. When customers defect, ComputeFast can try to shore up service while embarking on more expansion plans to maintain revenues. In the long run, however, these choices could lead to a reputation for poor service, which may necessitate

FIGURE 5.5

Pursuing Two Goals at Once

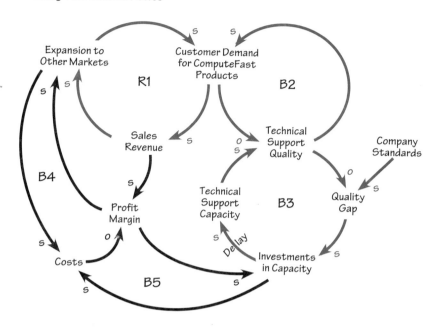

If investments in technical support capacity and overseas market expansion are pursued simultaneously, ComputeFast's profit margins will be squeezed (B4 and B5). Two possible alternatives are to pursue one goal at the expense of the other, or to accept lower profit margins in the short term as the company builds the necessary infrastructure to support both markets.

additional price cuts to regain customers. Price cuts, in turn, would squeeze margins and create pressure to improve them, perhaps by cutting back on service investments.

Here's an alternative, possibly better scenario: ComputeFast might commit to the necessary investments in tech support and overseas expansion plans, but expect lower profit margins in the short term. By relaxing the profit margin goal in the short term, the company may be better able to sustain healthy profit margins *in the long run,* once it has built the necessary infrastructure to support its new markets.

ComputeFast as a Complex System

The ComputeFast story displays several characteristics of a complex system:

- At the company's highest levels, there's an ever-present risk of conflict between the goals of expanding markets and the need to maintain adequate service quality.

- To compound this conflict, there's the persistent possibility that feedback about current service-quality levels will be distorted, delayed, or misinterpreted. On the other hand, the company has the power to modify its behavior if it chooses to gather and track as much feedback as possible about service quality.

- Service quality at ComputeFast has a tendency to drift back to its previous norms—a sign of one or more balancing processes at work.

- ComputeFast has the potential to replicate itself elsewhere in its environment, as well as modify its environment, if it succeeds in its expansion goals.

Understanding the special nuances of a complex system like ComputeFast and being able to diagram the system is not a "quick fix." However, if you were a manager at ComputeFast, you and your colleagues could use causal loop diagrams, like those shown in the story, to better see the dynamics of the company and design and test potential strategies. An even more effective exercise would be to use computer modeling to simulate and test the various scenarios—this kind of software lets you explore the impact of decisions much faster and more accurately than you can on paper. You would then have a better chance of anticipating the consequences of your choices and preparing yourself for the inevitable delays and side-effects of your policies.

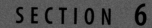

From Loops to Leverage:

Applying and Practicing Systems Thinking

N ow that you have some familiarity with interpreting and creating causal loop diagrams, and with complex systems and the idea of systems thinking overall, what next? Systems thinking tools are useful only if you can actually practice and apply them. Many people have compared the process of learning to think systemically to that of learning a new language—and no one expects to master Japanese by reading a single workbook! As with any new language, fluency in systems thinking requires lots of study and practice.

Don't get discouraged yet, however: There are many ways of applying systems thinking tools immediately—both individually and in your organization. In this final section, we offer an array of practical tips—from a list of "do's" and "don'ts" for systems thinking on the job, to ideas for making systems thinking an individual, life-long practice.

THE DO'S AND DON'TS OF SYSTEMS THINKING ON THE JOB[1]

Besides using BOT graphs and CLDs to generate insights into problems, how else can you begin using systems thinking on the job, especially if you want to introduce it to others? Your best bet is to approach this endeavor in the spirit of "learning to walk before you run." Here are some suggestions:

General Guidelines

DON'T use systems thinking to further your own agenda.
Systems thinking is most effective when it's used to look at a problem in a new way, not to advocate a predetermined solution. Strong advocacy will create resistance—both to your ideas, and to systems thinking itself. Present systems thinking in the spirit of inquiry, not inquisition.

DO use systems thinking to sift out major issues and factors.
Systems thinking can help you break through the clutter of everyday events to recognize general patterns of behavior and the structures that are producing them. It also helps in separating solutions from underlying problems. Too often we identify problems in terms of their solution; for example, "The problem is that we have too many _____ (fill in the blank: people, initiatives, steps in our process)," or "The problem is that we have too little _____ (resources, information, budget)."

DON'T use systems thinking to blame individuals.
Chronic unresolved problems are more often the result of systemic breakdowns than individual mistakes. Solutions to these problems lie at the systemic, not the individual, level.

DO use systems thinking to promote inquiry and challenge preconceived ideas.
Here are some clues that nonsystemic thinking is going on: phrases such as "We need to have immediate results," "We just have to do *more* of what we did last time," or "It's just a matter of trying harder."

Getting Started

DON'T attempt to solve a problem immediately.
Don't expect persistent and complex systemic problems to be represented, much less understood, overnight. The time and concentration required should be proportional to the difficulty and scope of the issues involved. Your goal should be to achieve a fuller and wider understanding of the problem.

---

1. This section was written by Michael Goodman and originally published in *The Systems Thinker*™ Newsletter, Volume 3, No. 6. Michael Goodman is a principal of Innovation Associates, an Arthur D. Little Company, and heads IA's Systems Thinking Group.

DO start with smaller-scale problems.

Don't attempt to diagram the whole system—otherwise you'll quickly become overwhelmed. Instead, try to focus on a problem issue and draw the minimum variables and loops you'll need to capture the problem.

DON'T work with systems thinking techniques under pressure, or in front of a group that is unprepared for or intolerant of the learning process.

If your audience is not familiar with the concepts and methods of systems thinking, they might not understand that the process reveals mental models, can be controversial, and is highly iterative in nature. It is far more beneficial to have the group engage in their own loop building after appropriate instruction and foundation have been given.

DO develop your diagrams gradually and informally, in order to build confidence in using systems thinking.

Look at newspaper articles and try to draw a few loops that capture the dynamics of a problem being described.

DON'T worry about drawing loops right away.

One of the strongest benefits of the systems thinking perspective is that it can help you learn to ask the right questions. This is an important first step toward understanding a problem.

Drawing Causal Loop Diagrams

DO start with the process of defining variables, and encourage airing of assumptions.

This approach leads to a better shared understanding of a problem. Diagramming is a very effective tool for promoting group inquiry into a problem or issue.

DO start with a central loop or process, then add loops to "fill in" detail.

For example, the central loop may show how the system is supposed to work, and the additional loops can explore what is pushing it out of whack.

DON'T get bogged down in details.

Start simply, at a high level of generalization, but with enough detail to sum up the observed behavior. For example, if you're exploring the causes of missed delivery dates in a factory, lump together the types of products that are experiencing similar delays.

DO begin by looking for generic structures that might clarify the problem.

Generic structures such as archetypes provide a focal point or a storyline to begin the process of understanding a problem.

DO work with one or more partners.

Multiple viewpoints add richness and detail to the understanding of a problem.

**DO check with others to see if they can add
some insight or improve upon your diagram.**
Especially consult people in other functional areas who might have a
different perspective on the problem. For example, with a manufacturing delay problem, you might check with finance to see whether there
are any dynamics in the finance arena that are affecting the manufacturing delays (capital investments and purchases, etc.). The same can be
done for marketing, sales, and so forth.

DO work iteratively.
There's no "final" or "correct" causal loop diagram. Looping is a learning
process that should continue to evolve with new data and perspectives.

DON'T present "final" loop diagrams as finished products.
Instead, present a diagram as a tentative and evolving picture of how you
are seeing things. To gain buy-in and maximize learning, make sure your
audience participates in the modeling process.

DO learn from history.
When possible, check data to see if your diagram correctly describes past
behavior.

Designing Interventions

DO get all stakeholders involved in the process.
This will help ensure that all viewpoints have been considered, and will
increase the chance that the stakeholders will accept the intervention.

DON'T go for vague, general, or open-ended solutions.
For example, instead of proposing a solution such as "Improve communications," rephrase your thought as "Reduce the information delay between
sales and manufacturing by creating a new information system."

**DO make an intervention specific, measurable,
and verifiable.**
For example, "Cut the information delay between sales and manufacturing
down to 24 hours."

DO look for potential unintended side effects of an intervention.
Remember the general principle: "Today's problems often come from yesterday's solutions." Any solution is bound to have trade-offs, so use systems
thinking to explore the implications of any proposed solution *before* trying
to implement it.

**DON'T be surprised if some situations defy solution,
especially if they are chronic problems.**
Rushing to action can thwart learning and ultimately undermine efforts to
identify higher leverage interventions. Resist the tendency to "solve" the
issue, and focus on gaining a deeper understanding of the structures producing the problem. Be wary of a symptomatic fix disguised as a long-term,
high-leverage intervention.

**DO remember that, even for systems thinkers,
it's easy to fall back into a linear process.**

Learning is a cycle—not a once-through process with a beginning and an end. Once you have designed and tested an intervention, it's time to shift into the active side of the learning cycle. This process includes taking action, seeing the results, and then coming back to examine the outcomes from a systemic perspective.

PRACTICING LIFE-LONG SYSTEMS THINKING[2]

In addition to using systems thinking skills in the workplace, how can you incorporate the learning and practice of systems thinking into everyday life? Happily, there are lots of opportunities to do this, not only as individuals but also as in collaborative learning arrangements and in learning communities.

Individual Practice

Individual practice is a good starting point for applying the basic concepts of systems thinking that you've learned through a book or a workshop. By incorporating some of the basic tools and understanding of the systems approach into everyday work situations, you can begin to build your confidence and competence—and gain a clearer sense of where you need further practice. These practice methods need no equipment—just your brain, your curiosity, and your enthusiasm. Here are some suggestions for everyday use of systems thinking skills:

Pay Attention to the Questions You Ask

Much of the value of systems thinking comes from the different framework that it gives us for looking at problems in new ways. For example, the concept of the Events-Patterns-Structure iceberg (Figure 6.1, "The Iceberg") lets us practice going beyond event-oriented responses to look for deeper, structural causes of problems.

FIGURE 6.1

The Iceberg

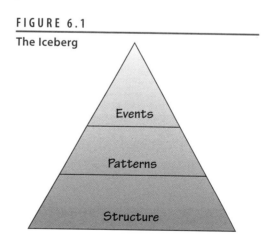

---

2. This section was written by Linda Booth Sweeney and originally published in *The Systems Thinker*™ Newsletter, Vol. 7, No. 8. Linda Booth Sweeney is an educator, researcher, and associate of the MIT Center for Organizational Learning. She is also the author of *The Systems Thinking Playbook, Volume I,* and co-author of *The Systems Thinking Playbook, Volume II.*

To practice moving from events to structure, start by simply paying attention to the questions you ask. Practice asking questions that get at deeper meanings, that inquire into others' viewpoints, or that elicit additional information. Examples include:

- Questions that look for patterns
  ("Has this same problem occurred in the past?");

- Genuinely curious questions that enable new information to be shared
  ("What makes you say that?");

- Questions that push for a deeper understanding of the problem
  ("What structures might be causing this behavior?");

- Questions that probe for time delays
  ("What effect will project delays have on our resources?");

- Questions that look for hidden loops
  ("What might be causing this feeling that we are 'spinning our wheels?'"); and

- Questions that look for unintended consequences
  ("What would happen if we got what we wanted?").

When probing for potential unintended consequences, the challenge is to ask the question without causing people to feel threatened or having yourself labeled as the "naysayer." One organization came up with a clever way to combat the negative labeling of those who raised concerns: The CEO and his senior team established a "qualming" period during each team meeting. During this time, team members were encouraged and expected to raise their concerns about the topic at hand. By setting aside a designated time for this activity, they turned the practice of looking for unintended consequences into a positive and creative part of the group's process.

Sense and Notice the Systems Around You
The great poet and philosopher Rumi once wrote: "New organs of perception come into being as a result of necessity. Therefore . . . increase your necessity so that you may increase your perception." We can increase our ability to perceive and sense the three levels of events-patterns-structures operating simultaneously by actively looking for feedback loops in everyday situations. How? We simply need to stop, look, listen, and sense.

For one colleague, the daily practice of thinking systemically comes with taking care of her house and garden, as she reflects on how her actions are affecting the Earth, the water, the air, and all the other living creatures. Little by little, that practice is leading her to use different cleaning agents, cut out most pesticides, let the clover take over the lawn, and plant lots of shrubs and trees to provide a rich, varied habitat. For her, the connection to systems thinking is to remember the Earth as a system, of which she is a part, and to try to imagine the impact of her actions on that larger system.

Draw a Loop-a-Day (or one a week)
Try instituting a new morning ritual: Sit down with your cup of coffee, the newspaper, a pad of paper, and a pen, and look for stories that can be explored through causal loop diagrams. Search for stories that describe pat-

terns of behavior over time ("The unemployment rate rose over the past 10 years, as did the number of families seeking welfare assistance.") and sketch out the systemic structure that you think might have produced those patterns of behavior. This is an excellent way to gain practice in recognizing systemic structures at work, and to master the mechanics of drawing causal loop diagrams.

Collaborative Learning

When it comes to practice, we are each other's greatest assets. You can greatly enhance your understanding and application of systems thinking by working in a group, in which you can offer feedback and learn from each other's experience. Here are several ideas for creating group processes that can further your systems thinking practice:

Apprenticeship/Mentoring/Coaching

If you can find the right person or organization, try to establish an apprenticeship with someone who is more skilled than you in the tools and techniques of systems thinking. As part of your "training," see if you can shadow that person during the workday (if he or she is a manager) or on their next engagement (if he or she is a consultant).

You can also identify someone to be your coach or mentor. Fill a large file folder with causal loops from newspaper and magazine articles and ask a colleague who is well versed in the field of system dynamics to act as your coach. Pick one example a week from your daily coffee-and-causal-loop exercise and fax your coach both the article and your causal loop diagram explanation. Then meet—even if by telephone—to discuss the loop and consider alternative scenarios and possible interventions.

Book Group

Another possibility is to find a partner or group with whom you can connect on a regular basis (perhaps monthly) to read a book on systems thinking or organizational learning. Perhaps you want to pick up a copy of a seminal book, such as *Limits to Growth* or *Principles of Systems*, and agree with a colleague to read a chapter a week, arranging a time to discuss that particular chapter (either in person, by phone, or via e-mail). You may also want to convene a small group to work through exercises in experientially based books such as *The Fifth Discipline Fieldbook* or *The Systems Thinking Playbook*—or even a workbook like this one! Even the president of a large nonprofit recently revealed that she and her staff meet once a week for several hours to work through a different systems thinking–related exercise every week.

Learning Communities

Learning communities involve individuals from cross-disciplinary backgrounds who come together to learn and practice in a group setting. There are currently several groups around the country that have formed to practice the skills of systems thinking, and the number continues to grow. In addition to face-to-face gatherings, several electronic forums have been created for people who want to exchange ideas and experiences in using the tools of systems thinking (see Appendix F: Additional Resources).

For example, to create a practice field for local practitioners and members of organizations interested in systems thinking, a colleague from Portland, Maine, organized monthly meetings at the houses of different group members. Group members brought stories from their business experiences—typically chronic problems that persisted despite ongoing efforts to resolve them. After several members told their "stories," one would be selected for the group to focus on. The rest of the meeting would be spent around a flip chart, exploring the roots of the problem through thoughtful inquiry and the construction of causal loop diagrams.

THE LEARNING JOURNEY

The most important underpinning of life-long systems thinking practice, whether in your personal or professional life, is an attitude of curiosity and mindfulness—that is, the ability to adopt a beginner's mindset and to maintain a heightened awareness of your thought processes. Most of the tips and practice opportunities described in this section provide a mirror that reflects your automatic responses in dealing with complex, systemic issues. Do you find yourself responding with openness and inquiry, or do you lapse into defensiveness and a need for "absolute" answers?

As you experiment with the various approaches to practicing and applying systems thinking, keep in mind that there is no one "right" way to master these new skills. These guidelines are simply intended to get you started. Over time, you may discover additional approaches that work well for you. For those adventurous souls who are attempting to take an unknown route to their destination, a lovely quote reminds us: "Traveler, there is no path. We make the road by walking."

Congratulations on working your way through this book! Now that you've had a taste of using systems thinking tools, we encourage you to continue your learning. There are lots of opportunities to practice systems thinking in your everyday life, and many resources for exploring the broad complement of systems thinking tools. We hope you found the information and activities in this workbook valuable. As we stated in the beginning, feel free to share your thoughts with us about what it was like to use this book, and about systems thinking overall. We look forward to hearing from you!

Additional Learning Activities

This appendix offers additional practice in using the tools of systems thinking. Some of these extra learning activities focus on the steps needed for uncovering systemic structures: formulating a problem, identifying variables, drawing BOTS, and creating CLDs. Others encourage you to stretch your thinking a little more by working with multiloop diagrams.

But don't stop here! As you saw in Section 6, there are lots of opportunities to practice systems thinking in your everyday life. We hope you'll take advantage of them and become a life-long systems thinker.

ACTIVITY 1 | **BUDGET BUGABOOS**

Purpose: To uncover the systemic structure at work in the story

Outcome: Insights into the systemic nature of the story
A behavior over time graph and causal loop diagram

Instructions: Read the story below and then answer the questions that follow.

➤ *Budget Bugaboos*

Budgets are great for tracking money as it flows through a company. But when they are used for other purposes—long-range planning or gauging performance—they can distort reality and mislead managers. When budgets become an end in themselves, the company suffers. The result: Managers end up managing the budget rather than managing the company.

Here's what typically happens: As a result of the annual budgeting process, each division is given a budget, a target level of spending. If actual costs exceed the budget, everyone in the division begins to experience budget pressure. A division may respond in a variety of ways: laying off employees, cutting back the work week, dropping "perks," and eliminating expenditures for training and outside conferences. Eventually, to management's relief, the monthly and quarterly results improve.

QUESTIONS

1. To formulate the problem, briefly summarize the above story.

2. What are the story's key variables? (Hint: Try to look for four.) List them below.

3. Graph these variables' behavior over time.

4. Draw a causal loop diagram that represents the variables' interrelationships. Be sure to indicate any important delays and to label each link with an s or an o. Label the overall loop with a B or an R for balancing or reinforcing.

5. Trace around your diagram, telling the story, to test the soundness of your diagram.

ACTIVITY 2   **MANAGING THE ELECTRIC COMPANY**

Purpose: To uncover the systemic structure at work in the story

Outcome: Insights into the systemic nature of the story
A behavior over time graph and causal loop diagram

Instructions: Read the story below and then answer the questions that follow.

➤   *Managing the Electric Company*

Like other members of the American electric utility industry, Statewide Power Company faced fundamental challenges to its core business. These challenges came from price pressures from large industrial customers, trends in deregulation, and state regulation.

For companies like Statewide, large industrial customers competing directly or indirectly in the global economy were pressuring electricity suppliers to match world standards of productivity and quality. By heightening expectations for continuous improvements in these areas, these customers drew power companies into the global competition of the world market.

The industries competing in the global market had traditionally worked through the state regulatory system to gain the electricity prices they needed to be competitive. For the most part, however, state regulation had not lowered prices enough, in these companies' opinion. Global competitors therefore advocated federal deregulation of electric utilities and open access for consumers to choose among competing electricity generators. In addition, national and international investment houses and the competitive business community saw the profit potential that could result from freeing large companies from regulation.

All these developments put pressure on Congress to pass federal laws to promote competition through the open access of existing electric utilities' transmission grids. This idea of electricity deregulation revealed a major shift in the way electric utilities and regulators think about how the system should operate, and the kinds of results the system should produce.

Despite federal interest in pursuing the kind of deregulation seen in the airline, telephone, and natural gas industries, the American electric utility industry remained one of the few regulated U.S. industries until recently. In 1992, Congress passed the Policy Energy Act, which promoted competition between electric utilities.

QUESTIONS

1. Those in favor of deregulation of the electric utilities have a theory about how competition can lower electricity prices. What is their theory?

2. To uncover the structure that underlies this theory, begin by listing the key variables in the theory. (Hint: Try to list four.)

3. Graph the behavior of these variables over time.

4. Now draw a causal loop diagram that shows how the variables influence each other. As with Activity 1, make sure each link in your diagram has an s or an o. Label your loop with an R or a B. Add any necessary delays. Finally, check the diagram's accuracy when you're finished.

ACTIVITY 3   **ADDICTED TO OIL**

Purpose: To uncover the systemic structure at work in the story

Outcome: Insights into the systemic nature of the story
A multiloop causal diagram

Instructions: Read the story below and then answer the questions that follow.

➤ *Addicted to Oil*

In an effort to ease periods of energy shortages, Americans since the mid-1980s have imported more and more barrels of oil to ensure their daily "fix." Unwilling as a country to restrict use of our cars and other luxuries, we have grown addicted to foreign oil supplies. The U.S. government has even engaged in a military buildup in the Middle East to secure this short-term source of oil.

At the same time, American scientists have tried to develop options for alternate energy sources. Switching from an oil-based economy to one based on multiple energy sources poses quite a challenge and requires a long-term commitment to the strategy. It is difficult to focus on developing alternative solutions when, every day, the country hungers for more and more oil and gives in to the temptation to buy foreign oil. As more attention is turned toward obtaining foreign oil for short-term satisfaction, less is invested in developing alternative energy sources.

QUESTIONS

1. To build a causal loop diagram of this story, start by identifying the key variable that leads to everything else. That is, what prompts Americans to buy foreign oil and to look for alternate sources of energy?

2. Without worrying about loops yet, think about the nature of this dynamic. As energy shortages increase, Americans buy more foreign oil and develop other energy resources, and then the energy shortage gets eased. What kind of process do you sense at work here: reinforcing or balancing?

3. Draw a simple multiloop diagram that shows how energy shortages, oil imports, and development of alternate energy sources influence each other.

4. Now think about the role that oil addiction—the craving for a short-term "fix"—plays in this scenario. Each time the U.S. buys foreign oil, it becomes more and more dependent on that energy source. As this addiction grows stronger, it diverts attention away from efforts to develop alternate energy sources. Neglecting alternate sources in turn only forces Americans to become even more reliant on imports. What kind of process does this part of the scenario sound like to you: reinforcing or balancing?

5. Try adding an arrow that links "Addiction to Oil Imports" to the diagram you drew in Step 3. Add any important delays.

6. What does the new version of your diagram suggest about the impact that an intense, short-term need can have on long-range, more effective solutions to the original problem of energy shortages?

ACTIVITY 4   THE NATIONAL ECONOMY

Purpose: To uncover the systemic structure at work in the story

Outcome: Insights into the systemic nature of the story
A behavior over time graph and causal loop diagram

Instructions: Read the story below (excerpted from Jay Forrester's description of the national economy as studied through the National Model), and then answer the questions that follow.

➤ *The National Economy*

When consumer demand increases, as it did in the 1950s, after World War II, production rises. With expanding production, the need for more workers means manufacturing wages rise. At the same time, increasing production leads manufacturers to want to make even more profits. They need to expand, building more factories, warehouses, and distribution facilities, and they need to invest in more labor and raw materials. They have an insatiable demand for capital. With optimism high, banks are willing to finance a growing stream of loans, which go into, among other things, wages for the workers who construct the new plant and build the new equipment. In general, wages for both construction and production workers boom. This newly affluent middle class of workers satisfies its desire for the "good life" by spending ever more on cars, appliances, clothing, and electronics.

QUESTIONS

1. Write a brief synopsis of the story.

2. Identify the story's key variables. (Hint: Try to list seven.)

3. Graph these variables' behavior over time.

4. Draw the causal loop diagram. (Hint: The diagram contains two loops of the same type.)

ACTIVITY 5   **THE RISING COST OF HEALTHCARE**

Purpose: To uncover the systemic structure at work in the story

Outcome: Insights into the systemic nature of the story
A behavior over time graph and causal loop diagram

Instructions: Read the story below and then answer the questions that follow.

➤   ***The Rising Cost of Healthcare***

In the business world, companies hoping to control the rising cost of health-care benefits have begun studying the ways in which healthcare services are used by employees. For a while, these studies helped businesses to restrict unnecessary care. One study, for example, analyzed several hundred insured groups over a three-year period and concluded that hospital admissions had decreased 13 percent. Overall, employers were able to cut medical costs 3 percent through using such studies.

However, concerns arose about the financial benefits of these strategies because of the heavy administrative burden they placed on doctors and hospitals. Indeed, some analysts estimated that as much as 20 percent of the cost of healthcare could be attributed to excess paperwork and other administrative tasks that befall providers. By the year 2020, they warned, such spending could amount to half of total healthcare dollars.

Studying healthcare use had become popular because the cost savings far outweighed the insurance fees charged to employers. Yet over the long term, the increased costs to healthcare providers would lead to higher costs for ser-vices, further shifting costs to patients, and—in the end—to employers. These increased costs were especially insidious because the extra time it takes for a physician to answer a phone call or chase down information for a report is not measured.

QUESTIONS

1. To uncover the system at work in this story, begin by looking again at the first paragraph. What two variables do you detect are being discussed in that paragraph?

2. Graph the two variables' behavior over time.

3. Now draw a simple loop that shows these variables' interrelationship. Is the loop reinforcing or balancing? Label it with an R or a B.

4. Reread the second and third paragraphs of the story. What two additional key variables do these paragraphs introduce?

5. Draw a new behavior over time graph that includes the variables you graphed in Step 2 and the additional variables you listed in Step 4.

6. Now draw a new CLD that incorporates all four variables of the story. (Hint: This will be a double-loop CLD.) Be sure all the links are labeled with an s or an o, and that each loop is labeled with an R or a B. Add any important delays.

Learning Activity Key Points and Suggested Responses

Before you become anxious comparing your responses to those in this section, remember that for many learning activities, there's no one right answer. We offer the following suggested responses as guidelines, to invite you to stretch your thinking about the various problems and scenarios presented in the learning activities. You may well come up with your own original insights into the activities. After all, an important benefit of using systems thinking tools is that they bring to the surface our assumptions about issues and problems and give us a starting place to address them.

Note: For learning activities that asked for very subjective answers, we have not listed any suggested responses.

SECTION 3 **UNCOVERING SYSTEMIC STRUCTURES: DRAWING BEHAVIOR OVER TIME GRAPHS**

ACTIVITY 1   *THE PROBLEM WITH PRICE PROMOTIONS*

1. *What are the problems described in the case?*

   Slowdown in U.S. population growth
   Smaller annual increases in consumption
   Promotions eroding brand image
   Promotions encouraging price-based shopping
   Manufacturers and retailers getting hooked on promotions
   Supermarkets' growing control over the promotions
   Supermarkets demanding subsidies
   Forward buying
   Diverted low-price goods

2. *From the point of view of the manufacturers, what is the overarching problem that includes many of the specific problems you named in Question 1?*

   Price promotions eroded brand image, encouraged consumers to shop on price only, and led supermarkets to misuse promotions. The use of price promotions was a problem.

3. *Is there an even deeper problem behind the one you named in Question 2? If so, what is it?*

   A deeper problem is that manufacturers are unable to create meaningful distinctions among their brands. An even deeper problem is the slowdown in population growth and the rate of consumption.

ACTIVITY 2   THE CASE OF THE ENERGY DRAIN

1. *What's the problem in this story?*

   Your first thought might be that the problem is the amount of coffee the narrator is consuming. But why does the narrator drink coffee? To wake up, to stay awake in the middle of the morning, and to perk up in the middle of the afternoon. From that perspective, you could say the problem is that the narrator keeps running out of energy.

2. *What are the three or four most important variables in the case?*

   Energy level
   Coffee or Drinking coffee or Use of coffee or caffeine
   Dependence on coffee or caffeine

3. *What is the behavior of those variables over time? Graph them.*

   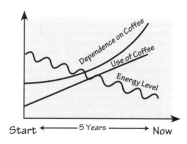

4. *What do you observe about the behavior of the variables?*

   In this scenario, the use of coffee rises steadily, while dependence on coffee rises sharply. Energy level fluctuates but declines overall.

5. *What relationships among the behavior patterns of the variables do you observe?*

   Use of coffee and dependence on coffee seem paired in this situation. Also, the more the use of coffee and dependence on it rises, the more energy level drops—exactly the opposite of what was desired!

ACTIVITY 3   THE CASE OF THE AUDIO-ELECTRONIC ROLLER COASTER

1. *What was the long-standing, chronic problem facing AudioMax Corp.?*

   The apparent problem is that after a period of steady growth, AudioMax Corp. suffered from shaky financial performance that started with a serious decline, and continued as a "boom and bust" pattern. AudioMax's uneven management quality was another problem. In fact, it looks as if the quality of the entire management team was uneven. A strong focus on technological improvements at the expense of other business processes could also lie at the root of the company's problems.

(Resist the temptation to conclude simply that the problem lay with AudioMax's CEO!)

2. *What two variables fed the initial steady growth in demand from AudioMax's customers?*

Technological innovation or Technological quality
New products

3. *Once AudioMax went public, what variable limited its ability to continue handling the growth of demand?*

Management strength, Management depth, or Management skill

(This variable stands for the issues related to failing to strengthen management at the time the company went public. It can include rising expenses and slumping productivity related to the new managers' lack of understanding of AudioMax's basic processes.)

4. *Graph the behavior over time of the two variables you identified in Question 2 and the one from Question 3. Then add the behavior over time of two more variables: "Demand" and "Ability to meet customer demand."*

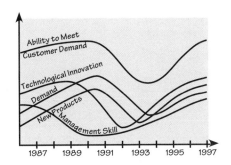

5. *What do you observe about the behavior of the individual variables over time?*

All the variables seem to rise steeply at first (except for Management skill, which dips early on). This is the growth phase. The variables then drop sharply (as the company's troubles begin), and finally start picking up again.

6. *What relationships do you observe among the variables?*

Management skill seems to drop first, then technological innovation, and finally the number of new products and ability to meet customer demand. After the management skills graph begins to rise again, so do all the other variables.

These relationships might lead you to hypothesize about the impact of management skill on success. For example, customer demand might be affected not just by marketing but also by the company's ability to deliver the product.

Section 4   UNCOVERING SYSTEMIC STRUCTURES: BUILDING CAUSAL LOOP DIAGRAMS

ACTIVITY 1   THE CASE OF THE PLATEAUING PROFITS

1. *Which kind of process—reinforcing or balancing—do you suspect was at play during Medicorp's **early** years?*

   Reinforcing, as indicated by the accelerated growth of Medicorp's customer base

2. *Which kind of dynamic—reinforcing or balancing—prevailed near the **end** of Medicorp's story?*

   Balancing, as indicated by the leveling off of the customer base

3. *Draw a simple BOT graph that shows just the pattern of the growth of Medicorp's customer base. Based on your graph, which kind of process—reinforcing or balancing—do you think describes the **overall** Medicorp story?*

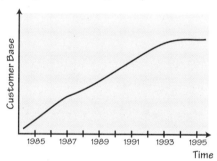

   This graph implies an overall balancing process, as the exponential growth eventually meets a limit and levels off.

4. *What do you think made Medicorp's growth hit a plateau?*

   Other HMOs began offering similar coverage policies and attracted business away from Medicorp, putting limits on its initial growth spurt.

ACTIVITY 2   THE CASE OF THE COLLAPSING BANKS

1. *What kind of dynamic—reinforcing or balancing—do you sense at work in the story?*

   This is a reinforcing process, recognizable by its accelerating change.

3. *What kind of loop is the collapsing-banks CLD? Label it with an R or a B.*

   The loop is a reinforcing process, and should be labeled with an R.

Activity 3   The "Organic To Go" Story

Part One

We have not provided suggested responses for the Part One "Organic To Go" questions; at this stage, the questions ask you simply to speculate on the company's situation.

Part Two

1. *Look again at the first Organic To Go causal loop diagram. Does it represent a reinforcing or a balancing process? Label the loop with an R or a B.*

   This is a reinforcing process, and the loop should be labeled with an R.

2. *Look at the second causal loop diagram. Does the new section of the diagram represent a reinforcing or balancing process? Label the section with an R or a B.*

   This is a balancing process, and the loop should be labeled with a B.

3. *When you look at both the original version and the expanded depiction of the story's actual outcome, do you see the issues at Organic To Go differently than you did at first? If so, how?*

   You might decide that the Organic To Go managers need to see beyond the initial "flush" of success that is depicted by the original reinforcing process of expansion. The high-energy growth during the company's early stages is not enough to sustain itself indefinitely.

4. *Given what you can see from the loop diagrams, what suggestions would you offer the managers?*

   Perhaps they could invest time and funds in more training for new employees in policies and expectations, before having them plunge into staffing the new stores. They might also think about slowing the pace of their growth a bit, to let the new staff get "up to speed." Attending to order filling, stocking, and maintenance problems is another idea. Finally, they could take steps to support newer managers and boost employee morale.

5. *What factors in particular do you now think the managers should pay most attention to?*

   Training of new employees and pace of overall growth are two important possibilities.

ACTIVITY 4   THE CASE OF THE RESTRICTED REVENUES

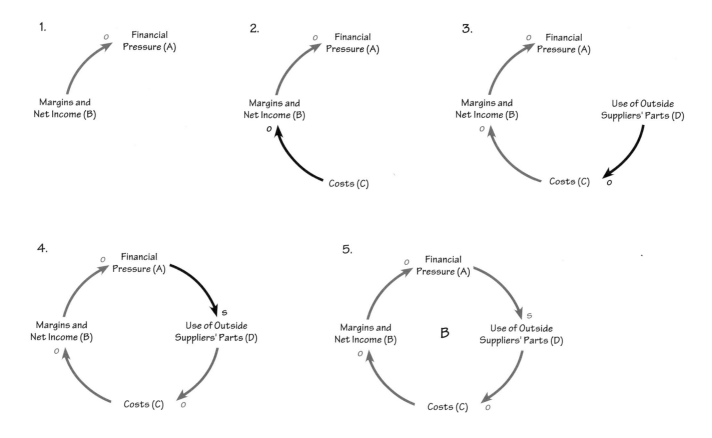

ACTIVITY 5   THE ALL-FOR-ONE COOPERATIVE

1. *Briefly sum up the All-for-One story.*

   As the All-for-One vision was built and reinforced, it encouraged a growing level of collaborative design among the members. Collaboration in designing the cooperative led members to work together on joint business experiments. The shared nature of these experiments seemed to encourage members to share with the rest of the cooperative what they learned, regardless of the success of the outcome. Sharing their learning and insight deepened the shared vision and reinforced the whole process.

2. *Identify four or five key variables from the story.*

   Shared vision
   Collaboration on design
   Joint experimentation
   Shared cost
   Shared learning and insight

3. *Graph the variables' behavior over time.*

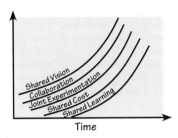

Time

4. *Decide how the variables interrelate, and draw a causal loop diagram to show the connections. Label the loop with an R or a B.*

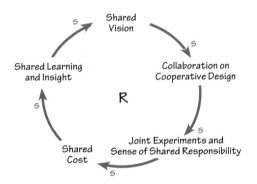

ACTIVITY 6   THE PROBLEM WITH USED CDs

1. *What was the original problem that CDs posed for record companies?*

   Because CDs cost a good deal more than LPs, record companies' original problem was how to persuade consumers to buy CDs.

2. *What did the record companies do to overcome this problem?*

   Their strategy was to promote CDs in various ways, including cheap deals through record clubs.

3. *What was the next problem or issue facing the record companies?*

   The next problem was a conflict with retailers over sales of used CDs. Record companies were concerned that sales of used CDs would cut into sales of higher priced, new CDs.

4. *Identify a total of four key variables reflecting the original problem and the later problem.*

   Record industry need to sell new CDs or Record industry pressure to sell new CDs
   Record industry CD promotions
   Retailer sales of new CDs
   Availability and sales of used CDs

5. *Create a graph of the variables' behavior over time.*

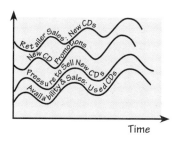

6. *Draw a simple two-loop diagram that shows the original problem and the later problem, including how the later problem is linked to the original problem. Label the two loops with an R or a B.*

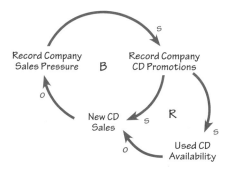

Here's the story that this CLD represents: There is pressure within the record industry to sell CDs. In response to this pressure, and to the need to overcome consumers' resistance, record companies run lots of low-price promotions. In the short term, this strategy produces sales of new CDs and reduces sales pressure. However, eventually the increasing number of new CDs in the hands of consumers leads to an increase in the sales of used CDs. Sales of used CDs then cut into the sales of new CDs, and sales pressure on the record companies goes up.

In this case, the short-term solution comes around again as a problem in the longer term.

APPENDIX A   ADDITIONAL LEARNING ACTIVITIES

ACTIVITY 1   BUDGET BUGABOOS

1. *To formulate the problem, briefly summarize the story.*

   If divisions' costs run over budget, the divisions feel pressure and respond by cutting costs. These measures eventually make results look better.

2. *What are the story's key variables?*

   Target budget
   Budget pressure
   Cost-cutting measures
   Costs

3. *Graph these variables' behavior over time.*

   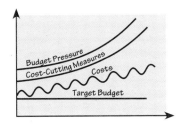

4. *Draw a causal loop diagram that represents the variables' interrelationships.*

   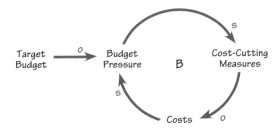

ACTIVITY 2   MANAGING THE ELECTRIC COMPANY

1. *Those in favor of deregulation of the electric utilities have a theory about how competition can lower electricity prices. What is their theory?*

   Advocates for deregulation want to use competition to lower electricity's unit prices. Those who promote deregulation feel that competition will stimulate increased productivity among energy suppliers, and that this increased productivity will be transferred to consumers through lower prices.

2. *To uncover the structure that underlies this story, begin by listing the key variables in the story.*

   Competition
   Productivity
   Unit electricity costs
   Electricity prices

3. *Graph the behavior of these variables over time.*

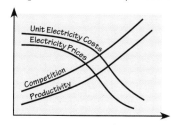

4. *Now draw a causal loop diagram that shows how the variables influence each other.*

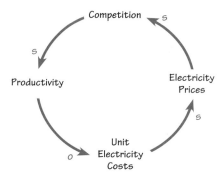

ACTIVITY 3   ADDICTED TO OIL

1. *To build a causal loop diagram of this story, start by identifying the key variable that leads to everything else. What prompts Americans to buy foreign oil and to look for alternate sources of energy?*

    Energy shortages

2. *What kind of process do you sense here: reinforcing or balancing?*

    Balancing

3. *Draw a simple multiloop diagram that shows how energy shortages, oil imports, and development of alternate energy sources influence each other.*

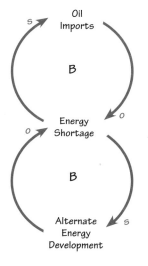

4. *Think about the role that oil addiction—the craving for a short-term "fix"—plays in this scenario. What kind of process does this part of the scenario sound like to you: reinforcing or balancing?*

Reinforcing

5. *Try adding an arrow that links "Addiction to Oil Imports" to the diagram you drew in Step 3. Add any important delays.*

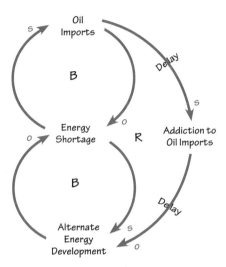

6. *What does the diagram suggest about the impact that an intense, short-term need can have on long-range, more effective solutions to the original problem of energy shortages?*

The Addiction to Oil Imports process is reinforcing, so over time it can take attention away from the more effective solution of developing alternate energy sources. By doing so, it only worsens the overall problem, because as efforts to develop alternate energy sources decrease, energy shortages rise even more, further adding to the temptation to buy foreign oil as a "quick fix."

ACTIVITY 4 THE NATIONAL ECONOMY

1. *Write a brief synopsis of the story.*

When consumer demand increases, production rises, leading to higher wages. At the same time, manufacturers seeking even higher profits desire to expand, and demand additional capital. Optimistic banks finance more loans, which provide capital and increase wages further. Wages for both construction and production workers boom, and these workers in turn spend more on luxuries.

2. *Identify the story's key variables.*

> Demand (consumer demand, consumer buying)
> Production (manufacturing)
> Need to expand or Demand for additional capital
> Borrowing (to buy capital plant) or Loans
> Wages for construction
> Wages for production (or manufacturing)
> All wages or Wages earned

3. *Graph these variables' behavior over time.*

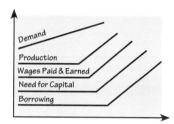

4. *Draw a causal loop diagram.*

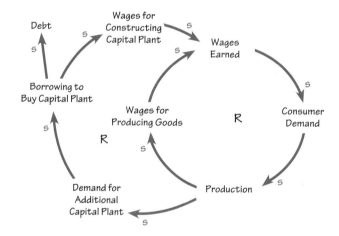

ACTIVITY 5   THE RISING COST OF HEALTHCARE

1. *Look again at the story's first paragraph. What two variables do you detect are being discussed in that paragraph?*

> Healthcare costs for businesses
> Use of healthcare studies

2. *Graph the two variables' behavior over time.*

3. *Draw a simple loop that shows these variables' interrelationship.*

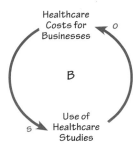

4. *Reread the second and third paragraphs of the story. What two additional key variables do these paragraphs introduce?*

Administrative burden on providers
Cost of administration

5. *Draw a new behavior over time graph that includes the variables you graphed in Step 2 and the additional variables you listed in Step 4.*

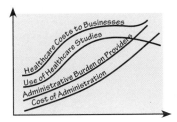

6. *Draw a new CLD that incorporates all four variables of the story.*

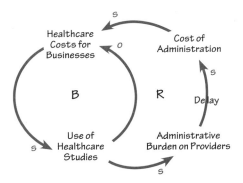

A Palette of Systems Thinking Tools

There is a full array of systems thinking tools that you can think of in the same way as a painter views colors—many shades can be created out of three primary colors, but having a full range of ready-made colors makes painting much easier.

There are a number of distinct types of systems thinking tools, all of which fall under several broad categories: dynamic thinking tools, structural thinking tools, and computer-based tools. Although each of the tools is designed to stand alone, they also build upon one another and can be used in combination to achieve deeper insights into dynamic behavior.

Dynamic Thinking Tools

Behavior Over Time Diagrams (BOTs) are more than simple line projections—they capture the dynamic relationships among variables. For example, say you wanted to project the relationship between sales, inventory, and production. If sales jump 20 percent, production cannot jump instantaneously to match the new sales number. In addition, inventory must drop below its previous level while production catches up with sales. By sketching out the behavior of different variables on the same graph, you can gain a more explicit understanding of how these variables interrelate.

Causal Loop Diagrams (CLDs) provide a useful way to represent dynamic interrelationships. CLDs make explicit your understanding of a system's structure, provide a visual representation to help communicate that understanding, and capture complex systems in a succinct form. CLDs can be combined with BOTs to form structure-behavior pairs, which provide a rich framework for describing complex dynamic phenomena. CLDs are the systems thinker's equivalent of the painter's primary colors.

Systems Archetypes is the name given to certain dynamics that seem to recur in many different settings. These archetypes, consisting of various combinations of balancing and reinforcing loops, are the systems thinker's "paint-by-numbers" set—you can take real-world examples and fit them into the appropriate archetype. They serve as a starting point from which you can build a clearer articulation of a business story or issue. The archetypes include "Drifting Goals," "Shifting the Burden," "Limits to Success," "Success to the Successful," "Fixes That Fail," "Tragedy of the Commons," "Growth and Underinvestment," and "Escalation."

DYNAMIC THINKING TOOLS

BEHAVIOR OVER TIME DIAGRAM

Can be used to graph the behavior of variables over time and gain insights into any inter-relationships between them. (BOT diagrams are also known as reference mode diagrams.)

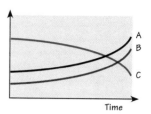

CAUSAL LOOP DIAGRAM

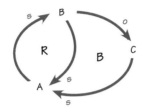

Used in conjunction with behavior over time diagrams, can help you identify reinforcing (R) and balancing (B) processes.

SYSTEMS ARCHETYPE

Helps you recognize common system behavior patterns such as "Drifting Goals," "Shifting the Burden," "Limits to Growth," "Fixes That Fail," and so on—all the compelling, recurring "stories" of organizational dynamics.

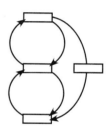

Structural Thinking Tools

Graphical Function Diagrams, Structure-Behavior Pairs, and *Policy Structure Diagrams* can be viewed as the building blocks for computer models. *Graphical Functions* are useful for clarifying nonlinear relationships between variables. They are particularly helpful for *quantifying* the effects of variables that are difficult to measure, such as morale or time pressure. *Structure-Behavior Pairs* link a specific structure with its corresponding behavior. *Policy Structure Diagrams* represent the processes that drive policies. In a sense, when we use these tools we are moving from painting on canvas to sculpting three-dimensional figures.

STRUCTURAL THINKING TOOLS

GRAPHICAL FUNCTION DIAGRAM

Captures the way in which one variable affects another, by plotting the relationship between the two over the full range of relevant values.

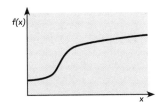

STRUCTURE-BEHAVIOR PAIR

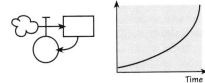

Consists of the basic dynamic structures that can serve as building blocks for developing computer models (for example, exponential growth, delays, smooths, S-shaped growth, oscillations, and so on).

POLICY STRUCTURE DIAGRAM

A conceptual map of the decision-making process embedded in the organization. Focuses on the factors that are weighed for each decision, and can be used to build a library of generic structures.

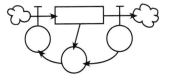

Computer-Based Tools

This class of tools, including *Computer Models, Management Flight Simulators,* and *Learning Laboratories,* demands the highest level of technical proficiency to create. On the other hand, very little advance training is required to use them once they are developed. These tools let you practice making decisions and observing the impact of those decisions—without actually risking your business. They also "collapse time"; that is, they let you see quickly how events would unfold over the long run as you implemented your policies in real life.

COMPUTER-BASED TOOLS

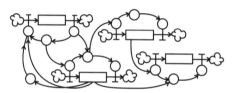

COMPUTER MODEL

Lets you translate all relationships identified as relevant into mathematical equations. You can then run policy analyses through multiple simulations.

MANAGEMENT FLIGHT SIMULATOR

Provides "flight training" for managers through the use of interactive computer games based on a computer model. Users can recognize long-term consequences of decisions by formulating strategies and making decisions based on those strategies.

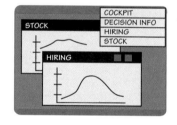

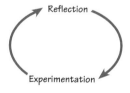

LEARNING LABORATORY

A manager's practice field. Is equivalent to a sports team's experience, which blends active experimentation with reflection and discussion. Uses all the systems thinking tools, from behavior over time diagrams to MFSs.

APPENDIX **D**

The Systems Archetypes

As a group, systems archetypes make up one of the 10 tools of systems thinking. The archetypes capture common "stories" that recur in different settings. They're valuable because they let you dig below the surface-level, distracting details of a complex situation to see the underlying systemic structure that drives a situation. Often, problems or issues that may seem unique at first can turn out to be caused by the same systemic structure, and therefore can be captured in the same systems archetype. The table below describes eight common archetypes and depicts their characteristic causal loop structures.

Drifting Goals

The "Drifting Goals" archetype states that a gap between a goal and an actual condition can be resolved in two ways: by taking corrective action to achieve the goal, or by lowering the goal. It hypothesizes that when there is a gap between the goal and the actual condition, the goal is lowered to close the gap. Over time, the continual lowering of the goal will lead to gradually deteriorating performance.

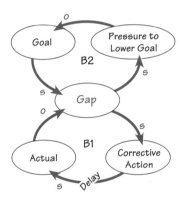

Escalation

The "Escalation" archetype occurs when one party's actions are perceived by another party to be a threat, and the second party responds in a similar manner, further increasing the threat. The archetype hypothesizes that the two balancing loops will create a reinforcing figure-8 effect, resulting in threatening actions by both parties that grow exponentially over time.

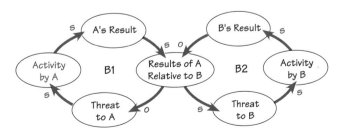

123

Fixes That Fail

The "Fixes That Fail" archetype states that a "quick-fix" solution can have unintended consequences that exacerbate the problem. It hypothesizes that the problem symptom will diminish for a short while and then return to its previous level, or become even worse over time.

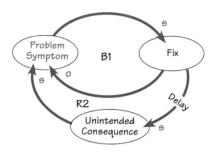

Growth and Underinvestment

The "Growth and Underinvestment" archetype applies when growth approaches a limit that can be overcome if capacity investments are made. If a system becomes stretched beyond its limit, however, it will compensate by lowering performance standards, which reduces the perceived need for capacity investments. This reduction also leads to lower performance, which further justifies underinvestment over time.

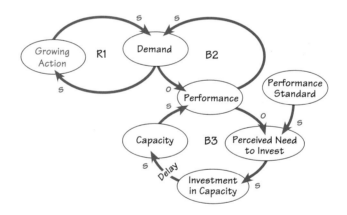

Limits to Success

The "Limits to Success" archetype states that a reinforcing process of accelerating growth (or expansion) will encounter a balancing process as the limit of that system is approached. The archetype hypothesizes that continuing efforts will produce diminishing returns as one approaches the limit.

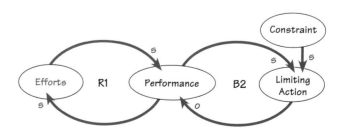

Shifting the Burden

The "Shifting the Burden" archetype states that a problem symptom can be resolved either by using a symptomatic solution or applying a fundamental solution. The archetype hypothesizes that once a symptomatic solution is used, it alleviates the problem symptom and reduces pressure to implement a more fundamental solution. The symptomatic solution also produces a side effect that systematically undermines the ability to develop a fundamental solution or capability.

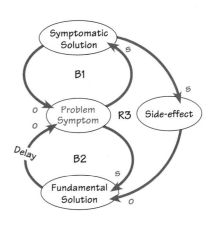

Success to the Successful

The "Success to the Successful" archetype states that if one person or group (A) is given more resources than another equally capable group (B), A has a higher likelihood of succeeding. The archetype hypothesizes that A's initial success justifies devoting more resources to A, further widening the performance gap between the two groups over time.

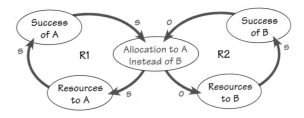

Tragedy of the Commons

The "Tragedy of the Commons" archetype identifies the causal connections between individual actions and the collective results (in a closed system). It hypothesizes that if the total usage of a common resource becomes too great for the system to support, the commons will become overloaded or depleted, and everyone will experience diminishing benefits.

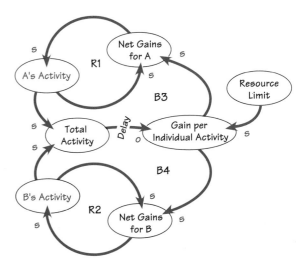

A Glossary of Systems Thinking Terms

Systems thinking can serve as a language for communicating about complexity and interdependencies. To be fully conversant in any language, you must gain some mastery of the vocabulary, especially the phrases and idioms unique to that language. This glossary lists many terms that may come in handy when you're faced with a systems problem.

Accumulator Anything that builds up or dwindles; for example, water in a bathtub, savings in a bank account, inventory in a warehouse. In modeling software, a stock is often used as a generic symbol for accumulators. Also known as **Stock** or **Level.**

Balancing Process/Loop Combined with reinforcing loops, balancing processes form the building blocks of dynamic systems. Balancing processes seek equilibrium: They try to bring things to a desired state and keep them there. They also limit and constrain change generated by reinforcing processes. A balancing loop in a causal loop diagram depicts a balancing process.

Balancing Process with Delay A commonly occurring structure. When a balancing process has a long delay, the usual response is to *over*correct. Overcorrection leads to wild swings in behavior. Example: real estate cycles.

Behavior Over Time (BOT) Diagram One of the 10 tools of systems thinking. BOT diagrams capture the history or trend of one or more variables over time. By sketching several variables on one graph, you can gain an explicit understanding of how they interact over time. Also called **Reference Mode.**

Causal Loop Diagram (CLD) One of the 10 tools of systems thinking. Causal loop diagrams capture how variables in a system are interrelated. A CLD takes the form of a closed loop that depicts cause-and-effect linkages.

Drifting Goals A systems archetype. In a "Drifting Goals" scenario, a gradual downward slide in performance goals goes unnoticed, threatening the long-term future of the system or organization. Example: lengthening delivery delays.

Escalation A systems archetype. In the "Escalation" archetype, two parties compete for superiority in an arena. As one party's actions put it ahead, the other party "retaliates" by increasing its actions. The result is a continual ratcheting up of activity on both sides. Examples: price battles, the Cold War.

Feedback The return of information about the status of a process. Example: annual performance reviews return information to an employee about the quality of his or her work.

Fixes That Fail   A systems archetype. In a "Fixes That Fail" situation, a fix is applied to a problem and has immediate positive results. However, the fix also has unforeseen long-term consequences that eventually worsen the problem. Also known as "Fixes That Backfire."

Flow   The amount of change something undergoes during a particular unit of time. Example: the amount of water that flows out of a bathtub each minute, or the amount of interest earned in a savings account each month. Also called a **Rate**.

Generic Structures   Structures that can be generalized across many different settings because the underlying relationships are fundamentally the same. Systems archetypes are a class of generic structures.

Graphical Function Diagram (GFD)   One of the 10 tools of systems thinking. GFDs show how one variable, such as delivery delays, interacts with another, such as sales, by plotting the relationship between the two over the entire range of relevant values. The resulting diagram is a concise hypothesis of how the two variables interrelate. Also called **Table Function**.

Growth and Underinvestment   A systems archetype. In this situation, resource investments in a growing area are not made, owing to short-term pressures. As growth begins to stall because of lack of resources, there is less incentive for adding capacity, and growth slows even further.

Learning Laboratory   One of the 10 tools of systems thinking. A learning laboratory embeds a management flight simulator in a learning environment. Groups of managers use a combination of systems thinking tools to explore the dynamics of a particular system and inquire into their own understanding of that system. Learning labs serve as a manager's practice field.

Level   See **Accumulator**.

Leverage Point   An area where small change can yield large improvements in a system.

Limits to Success   A systems archetype. In a "Limits to Success" scenario, a company or product line grows rapidly at first, but eventually begins to slow or even decline. The reason is that the system has hit some limit—capacity constraints, resource limits, market saturation, etc.—that is inhibiting further growth. Also called "Limits to Growth."

Management Flight Simulator (MFS)   One of the 10 tools of systems thinking. Similar to a pilot's flight simulator, an MFS allows managers to test the outcome of different policies and decisions without "crashing and burning" real companies. An MFS is based on a system dynamics computer model that has been changed into an interactive decision-making simulator through the use of a user interface.

Policy Structure Diagram   One of the 10 tools of systems thinking. Policy structure diagrams are used to create a conceptual "map" of the decision-making process that is embedded in an organization. It highlights the factors that are weighed at each decision point.

Rate See **Flow.**

Reference Mode See **Behavior Over Time Diagram.**

Reinforcing Process/Loop Along with balancing loops, reinforcing loops form the building blocks of dynamic systems. Reinforcing processes compound change in one direction with even more change in that same direction. As such, they generate both growth and collapse. A reinforcing loop in a causal loop diagram depicts a reinforcing process. Also known as vicious cycles or virtuous cycles.

Shifting the Burden A systems archetype. In a "Shifting the Burden" situation, a short-term solution is tried that successfully solves an ongoing problem. As the solution is used over and over again, it takes attention away from more fundamental, enduring solutions. Over time, the ability to apply a fundamental solution may decrease, resulting in more and more reliance on the symptomatic solution. Examples: drug and alcohol dependency.

Shifting the Burden to the Intervener A special case of the "Shifting the Burden" systems archetype that occurs when an intervener is brought in to help solve an ongoing problem. Over time, as the intervener successfully handles the problem, the people within the system become less capable of solving the problem themselves. They become even more dependent on the intervener. Example: ongoing use of outside consultants.

Simulation Model One of the 10 tools of systems thinking. A computer model that lets you map the relationships that are important to a problem or an issue and then simulate the interaction of those variables over time.

Stock See **Accumulator.**

Structural Diagram Draws out the accumulators and flows in a system, giving an overview of the major structural elements that produce the system's behavior. Also called flow diagram or accumulator/flow diagram.

Structure-Behavior Pair One of the 10 tools of systems thinking. A structure-behavior pair consists of a structural representation of a business issue, using accumulators and flows, and the corresponding behavior over time (BOT) diagram for the issue being studied.

Structure The manner in which a system's elements are organized or interrelated. The structure of an organization, for example, could include not only the organizational chart but also incentive systems, information flows, and interpersonal interactions.

Success to the Successful A systems archetype. In a "Success to the Successful" situation, two activities compete for a common but limited resource. The activity that is *initially* more successful is consistently given more resources, allowing it to succeed even more. At the same time, the activity that is *initially* less successful becomes starved for resources and eventually dies out. Example: the QWERTY layout of typewriter keyboards.

System Dynamics A field of study that includes a methodology for constructing computer simulation models to achieve better understanding of social and corporate systems. It draws on organizational studies, behavioral decision theory, and engineering to provide a theoretical and empirical base for structuring the relationships in complex systems.

System A group of interacting, interrelated, or interdependent elements forming a complex whole. Almost always defined with respect to a specific purpose within a larger system. Example: An R&D department is a system that has a purpose in the context of the larger organization.

Systems Archetypes One of the 10 tools of systems thinking. Systems archetypes are the "classic stories" in systems thinking—common patterns and structures that occur repeatedly in different settings.

Systems Thinking A school of thought that focuses on recognizing the interconnections between the parts of a system and synthesizing them into a unified view of the whole.

Table Function See **Graphical Function Diagram.**

Template A tool used to identify systems archetypes. To use a template, you fill in the blank variables in causal loop diagrams.

Tragedy of the Commons A systems archetype. In a "Tragedy of the Commons" scenario, a shared resource becomes overburdened as each person in the system uses more and more of the resource for individual gain. Eventually, the resource dwindles or is wiped out, resulting in lower gains for everyone involved. Example: the Greenhouse Effect.

The above glossary is a compilation of definitions from many sources, including:
- Innovation Associates' and GKA's Introduction to Systems Thinking coursebooks
- *The Fifth Discipline: The Art and Practice of the Learning Organization,* by Peter Senge
- High Performance Systems' *Academic User's Guide to STELLA*
- *The American Heritage Dictionary* and *The Random House Dictionary.*

Additional Resources

Newsletters

The Systems Thinker™ (Pegasus Communications)

Books

The Fifth Discipline: The Art and Practice of the Learning Organization, Peter M. Senge (Doubleday, 1990)

The Fifth Discipline Fieldbook, Peter Senge et al. (Doubleday, 1994)

The Systems Thinking Playbook, Linda Booth Sweeney and Dennis Meadows (The Turning Point Foundation, 1996)

Billibonk and the Thorn Patch, Philip Ramsey (Pegasus Communications, 1997)

Modeling for Learning Organizations, John D. W. Morecroft and John Sterman (Productivity Press, 1994)

Introduction to System Dynamics Modeling with DYNAMO, George P. Richardson and Alexander L. Pugh III (Productivity Press, 1981)

Introduction to Computer Simulation: A System Dynamics Modeling Approach, Nancy Roberts et al. (Productivity Press, 1983)

Short Volumes

A Beginner's Guide to Systems Thinking, Colleen Lannon (Pegasus Communications, 1990)

Systems One, Draper L. Kauffman, Jr. (Future Systems, Inc., 1980)

Systems Archetypes I: Diagnosing Systemic Issues and Designing High-Leverage Interventions, Daniel H. Kim (Pegasus Communications, 1992)

Systems Archetypes II: Using Systems Archetypes to Take Effective Action, Daniel H. Kim (Pegasus Communications, 1994)

Systems Thinking Tools: A User's Reference Guide, Daniel H. Kim (Pegasus Communications, 1994)

Applying Systems Archetypes, Daniel H. Kim and Colleen Lannon (Pegasus Communications, 1996)

Designing a Systems Thinking Intervention: A Strategy for Leveraging Change, Michael Goodman et al. (Pegasus Communications, 1997)

From Mechanistic to Social Systemic Thinking: A Digest of a Talk by Russell L. Ackoff, Lauren Johnson (Pegasus Communications, 1996)

Toward Learning Organizations: Integrating Total Quality Control and Systems Thinking, Daniel H. Kim (Pegasus Communications, 1997)

The Tale of Windfall Abbey, Margaret Welbank (BP Exploration Operating Company Limited, 1992)

Laminated Reference Guides

Systems Archetypes at a Glance (Pegasus Communications)

A Pocket Guide to Using the Archetypes (Pegasus Communications)

Electronic Learning Communities

Learning-org is an ongoing dialogue on organizational learning and the learning disciplines that is conducted on the Internet (via e-mail or the World Wide Web). For further information, contact the host, Richard Karash <rkarash@karash.com> or fax 617-523-3839.

Audiotapes and Videotapes

From Mechanistic to Social Systemic Thinking, Russell L. Ackoff (Pegasus Communications, 1993)

Designing Corporations for Success in the 21st Century, Jay Forrester (Pegasus Communications, 1995)

System Dynamics: The Foundation of the Learning Organization, John Sterman (Pegasus Communications, 1994)

Introduction to Systems Thinking: An Approach to Organizational Learning, David Kreutzer (Pegasus Communications, 1995) Audiotape only.

Simulation Software

ithink® 4.0, High Performance Systems, Hanover, NH

STELLA® 4.0, High Performance Systems, Hanover, NH

Vensim® Standard, Ventana Systems, Inc., Belmont, MA

Powersim® 2.5 Professional, Powersim Corp., Herndon, VA

Management Flight Simulators

People Express Management Flight Simulator™

The Beefeater™ Restaurants Microworld

Service Quality Microworld™

(all available through MicroWorlds, Cambridge, MA)

PEGASUS COMMUNICATIONS, INC. is dedicated to helping organizations soar to new heights of excellence. By providing the forum and resources, Pegasus helps managers articulate, explore, and understand the challenges they face in the complex, changing business world. For information about *The Systems Thinker*™ Newsletter, the annual *Systems Thinking in Action*™ Conference, the *Power of Systems Thinking*™ Conference, or other publications that are part of *The Organizational Learning Resource Library*™ Catalog, contact:

PEGASUS COMMUNICATIONS, INC.
PO Box 120   Kendall Square
Cambridge, MA   02142-0001   USA

Phone: (617) 576-1231
Fax: (617) 576-3114
Web Site: http://www.pegasuscom.com